THE FORGOTTEN TRIBE: SCIENTISTS AS WRITERS

PERSPECTIVES ON WRITING
Series Editors, Susan H. McLeod and Rich Rice

The Perspectives on Writing series addresses writing studies in a broad sense. Consistent with the wide ranging approaches characteristic of teaching and scholarship in writing across the curriculum, the series presents works that take divergent perspectives on working as a writer, teaching writing, administering writing programs, and studying writing in its various forms.

The WAC Clearinghouse, Colorado State University Open Press, and University Press of Colorado are collaborating so that these books will be widely available through free digital distribution and low-cost print editions. The publishers and the Series editors are committed to the principle that knowledge should freely circulate. We see the opportunities that new technologies have for further democratizing knowledge. And we see that to share the power of writing is to share the means for all to articulate their needs, interest, and learning into the great experiment of literacy.

Recent Books in the Series

Jacob S. Blumner and Pamela B. Childers, *WAC Partnerships Between Secondary and Postsecondary Institutions* (2015)

Nathan Shepley, *Placing the History of College Writing: Stories from the Incomplete Archive* (2015)

Asao B. Inoue, *Antiracist Writing Assessment Ecologies: An Approach to Teaching and Assessing Writing for a Socially Just Future* (2015)

Theresa Lillis, Kathy Harrington, Mary R. Lea, and Sally Mitchell (Eds.), *Working with Academic Literacies: Case Studies Towards Transformative Practice* (2015)

Beth L. Hewett and Kevin Eric DePew (Eds.), *Foundational Practices of Online Writing Instruction* (2015)

Christy I. Wenger, *Yoga Minds, Writing Bodies: Contemplative Writing Pedagogy* (2015)

Sarah Allen, *Beyond Argument: Essaying as a Practice of (Ex)Change* (2015)

Steven J. Corbett, *Beyond Dichotomy: Synergizing Writing Center and Classroom Pedagogies* (2015)

Tara Roeder and Roseanne Gatto (Eds.), *Critical Expressivism: Theory and Practice in the Composition Classroom* (2014)

Terry Myers Zawacki and Michelle Cox (Eds), *WAC and Second-Language Writers: Research Towards Linguistically and Culturally Inclusive Programs and Practices,* (2014)

THE FORGOTTEN TRIBE: SCIENTISTS AS WRITERS

Lisa Emerson

The WAC Clearinghouse
wac.colostate.edu
Fort Collins, Colorado

University Press of Colorado
upcolorado.com
Boulder, Colorado

The WAC Clearinghouse, Fort Collins, Colorado 80523-1040

University Press of Colorado, Boulder, Colorado 80303

© 2017 by Lisa Emerson. This work is licensed under a Creative Commons Attribution-Non-Commercial-NoDerivatives 4.0 International.

Names: Emerson, Lisa, author.
Title: The forgotten tribe : scientists as writers / Lisa Emerson.
Other titles: Perspectives on writing (Fort Collins, Colo.)
Description: Fort Collins, Colorado : University Press of Colorado, [2016] | Series: Perspectives on writing | Includes bibliographical references.
Identifiers: LCCN 2016045184 | ISBN 9781607326434 (pbk.) | ISBN 9781607326441 (ebook)
Subjects: LCSH: Technical writing. | Academic writing. | Scientists' writings—Anecdotes. | Science writers—Anecdotes. | Literature and science.
Classification: LCC T11 .E545 2016 | DDC 808.06/65—dc23
LC record available at https://lccn.loc.gov/2016045184

Copyeditor: Julia Smith
Designer: Mike Palmquist
Series Editors: Susan H. McLeod and Rich Rice

The WAC Clearinghouse supports teachers of writing across the disciplines. Hosted by Colorado State University, and supported by the Colorado State University Open Press, it brings together scholarly journals and book series as well as resources for teachers who use writing in their courses. This book is available in digital formats for free download at wac.colostate.edu.

Founded in 1965, the University Press of Colorado is a nonprofit cooperative publishing enterprise supported, in part, by Adams State University, Colorado State University, Fort Lewis College, Metropolitan State University of Denver, Regis University, University of Colorado, University of Northern Colorado, Utah State University, and Western State Colorado University. For more information, visit upcolorado.com.

For my parents, Jean and Ellis Emerson,
who encouraged me to ask questions and pay attention.

CONTENTS

Prologue. A Book of Stories and Storytelling 3
Chapter 1. Countries Not Often Heard From. 7
Chapter 2. Reaching Out. 39
Chapter 3. The Reluctant Writers. 75
Chapter 4. The Writing Community 103
Chapter 5. The Development of the Scientific Writer. 129
Chapter 6. The Poets 155
Chapter 7. "We have to communicate the beauty and the passion." 179
Afterword. 215
Notes 217
References 219
Appendix A. Questionnaire for the Senior and Emerging Scientists 231
Appendix B. Interview Schedule for Senior and Emerging Scientists 233
Appendix C. Interview for Ph.D. Students. 237
Appendix D. The Scale Used to Develop the Quantitative Data. 241

THE FORGOTTEN TRIBE:
SCIENTISTS AS WRITERS

PROLOGUE
A BOOK OF STORIES AND STORYTELLING

I have long been interested in voices and conveying voices. I first encountered Mary Loudon's *Unveiled* (1992) almost 20 years ago. In her unusual book, Loudon interviews ten women in religious orders about the details of their lives and their perceptions of their roles, and then presents the interviews as uninterrupted extended narratives. I was struck by the way the distinctive voices of the women were an essential part of conveying their experience, that *what* they said was inextricable from *how* they said it. I was also intrigued by Loudon's technique of transforming the interviews into narratives, so that the voices were allowed to speak without the hindrance of the interviewer.

My interest in this form of narrative, constructed from oral interviews, deepened when I began to read Studs Terkel's remarkable set of work, which captures the voices of ordinary Americans on a range of topics, such as race (1992), work (1974), experience of WWII (1997), and death and dying (2014). Again, I was fascinated, not only by the content of these books, but by the immediacy of the narrative. "Listening" to voices captured on the page, some long silenced, conveyed both the weight and urgency of human experience in all its diversity. I have read and reread Terkel's books, as I have Mary Loudon's, for the sheer joy of experiencing the sound of the human voice and the variety of human experience.

Later, in the late 1990s, I encountered the ethical dilemmas of conveying voice in research, when I was writing up my study of scientists as teachers of writing (Emerson et al., 2000; Emerson, 2004). As I worked to integrate the voices of my interviewees into my analysis, using snippets of quotations, I felt some regret—and some ethical unease—that their voices were so muted through my own words, my own selection and analysis. I felt that I was short-changing my subjects—all of whom had shared their experiences and perceptions so richly and generously with me, several of whom had written huge volumes of reflection for my research—in restricting, selecting and editing their sustained and distinctive voices to my own purposes. I began investigating the process of developing narratives from interviews, and in 2003 developed a senior seminar on Life Writing where students interviewed family members and developed those interviews as narratives. The stories, and the voices that emerged, were rich, varied, and compelling stories of New Zealand life.

Prologue

My primary field of research, however, is science writing and writers, and during my career I have been fortunate to spend much of my days talking to scientists about writing. Over the years, I have valued these conversations, and enjoyed the subtlety with which many of my colleagues in the STEM disciplines talked about writing and their commitment to engaging with multiple audiences in deeply nuanced ways. As a writer and writing teacher, I have learnt much from these conversations.

Meantime, I was working with undergraduate science students, most of whom articulated reluctance about writing. The students in my freshman class almost universally disliked writing, claimed to have avoided writing where possible in school, and saw writing as completely unrelated to science. My senior class could see the relevance of writing to science but were (again almost universally) convinced they couldn't write.

I became intrigued. How did the students, with all their anxieties and fears about writing, transition into writing? How did they develop and transform into the sophisticated writers I observed amongst my colleagues?

When I turned to the literature to find answers to this question, I found very little about how scientists develop as writers beyond the undergraduate classroom. In particular, the perspectives of scientists seemed poorly—inadequately—represented in the literature. I began to feel a real concern: why were scientists not speaking into the scholarly discussions about scientific writing? Why were we not hearing their voices?

I began to collect these voices, and during this period, in 2010, attended the WRAB conference in Washington. At one session, following a paper on how scientists construct argument, the general discussion focused around scientists' inadequacies as writers; the general view seemed to be that scientists were poor writers, unnecessarily opaque, not interested in writing, and in need of remediation. Neither my experience nor my research supported this view, although this was certainly not the first time I'd heard such views. As I reflected on this discussion in the following weeks and months, it seemed to me that we had engaged in a form of cultural appropriation. I now saw the problem in a new way: it was not only important that we ask scientists to speak about writing, but we needed to somehow convey those voices in a way that allowed their voices to speak beyond the cultural appropriation of a scholarly humanities-based text. We needed to find a way for people to listen.

These factors, then—an interest in conveying "voices," a concern that scientists' voices are not sufficiently heard in the discussions about writing science and teaching science writing, a concern to avoid cultural appropriation, and a sense of our need to see the centrality of writing to scientists' professional lives—influenced the construction of this book, and in particular the decision to present sustained narratives as the primary driver of the text.

GIVING SOMETHING BACK

Collecting the voices for this book has been a rich and resonant experience, and I am so very grateful to all those who made themselves available, who invited me into their offices to talk about writing. It feels like a remarkable privilege to sit in someone's office and listen to their thoughts, and I have been almost always surprised by the generosity of the people I interviewed, by their willingness to explore territory they may not have thought about before—and often by their vulnerability and humility.

I have been delighted by the variety of stories and people I've engaged with. In one week alone, I listened to a research chemist who is a competitive body builder and a reader of romance fiction, talk with courage about her commitment to growth and learning as a writer. I sat in the book-lined office of a remarkably versatile writer of physics and history while he described decisions he had made as he wrote his book on quantum physics—the opening pages of which made me laugh out loud. I asked preposterous questions of a young mathematician (*"do you think in numbers and figures or words when you're thinking about math?"*), and was honoured by the care with which he explored possible answers, and enjoyed his laughter as he came to unexpected conclusions. I listened to a mathematician who also moonlights as a jazz musician talk about the differences between *writing up* and *writing down,* and an eminent physicist describe the fun of collaborating with a friend while hiking or ice fishing. He talked about how the science weaves in and out of their conversation, but how they always come back to the physics "because that's what we like talking about most."

My hope is that I've also given something back. Many times, after the tape recorder was switched off and we were talking more generally, the person I'd interviewed would say something along the lines of "you know, I didn't know I thought all that, I've never talked about writing before—it's been really valuable to think this out." Participants in this study have often emailed me after the interview—sometimes weeks after the interview—to share ideas they've been working on that were triggered by the interview, or to tell me how this interview has changed their thinking about their writing or changed the way they teach or work with graduate students or support their postdocs.

I hope, too, that by sharing these stories with teachers of writing, I'm giving something back to the scientific community. I'm hoping that some of those anecdotal views about whether scientists care about writing can, at the very least, be opened for examination.

Prologue

TELLING A STORY

I realise this is not a typical scholarly book—and this is intentional. Partly this is to do with my commitment not to engage in the cultural appropriation of these voices, though I must acknowledge, of course, the dialogic nature of the interviews, and that my background as a scholar in the humanities will inevitably have influenced both the construction of the narratives (in the questions I asked and did not ask, and in the process of editing the narratives), and the way I have chosen and arranged these narratives. This will be discussed further in Chapter 1.

But choices relating to the construction and "voice" of the book are also related to my aims and intended audience. My hope is that this book may be of interest and use, not just to writing scholars and teachers, but also to scientists and emerging scientists, as they strive to engage with their own writing and the writing of the scientific community. For this reason, while I have engaged with the literature, I have nevertheless tried to avoid using language that this second audience might find inaccessible.

I've used a personal writing style throughout this text because this book is about voices, and my voice as a teacher and researcher is a part of the story of this book. My motivation for writing was driven by personal experience: I have selected the participants and engaged with them on a personal level. To hide my own voice—even though I have aimed to mute it in the arrangement of the narratives—seemed to me disingenuous, and to disguise a significant component of the text.

One of the people I interviewed for this book told me that his favourite book, which he read over and over again, was Robert Browning's *The Ring and the Book*. A massive Victorian potboiler of a poem—it was a bestseller in its time—and a mystery thriller, *The Ring and the Book* tells the story of a particular event twelve times, from the point of view of different characters (some central, some tangential), not so that the one "true story" can be discerned but so that the variety of human perception can be portrayed (Slinn, 1991). This study tells the same story from multiple perspectives in the hope that readers will grasp, in a new way, the breadth of experience and perceptions of scientists as writers.

CHAPTER 1
COUNTRIES NOT OFTEN HEARD FROM

> I was interested in other countries not often heard from.
> — Studs Terkel, *Working*

I have argued elsewhere (Emerson, 2012) that scientists are, to a large extent, a lost or forgotten "tribe"[1] of academic writers—that they are, in Terkel's words (1997), a country "not often heard from." This is, of course, something of an exaggeration. Researchers from the humanities may watch scientists from a distance through the body of literature in the field of the rhetoric of science, which examines their cultural artefacts, that is, their textual products (Montgomery, 1996). Sometimes we get a little closer. Like early anthropologists who visit the tribal village, we might get as far as the lab to observe and document their textual practices and the social and cultural contexts in which these practices take place (Bazerman 1988, 1998; Doody, 2015; Graves, 2005).

But we know little of how scientists think as writers—about their beliefs, attitudes and experiences of writing—even though learning to adopt the attitudes and beliefs of their seniors will be essential to our science students in the future as they develop both their professional identity and their fluency as disciplinary writers (Beaufort, 2007; Dall'Alba & Sandberg, 2006; Gee, 2005; Harding & Hare, 2000; Poe et al., 2010). Learning to write as a scientist involves far more than imitating behaviour; for example, as many authors have shown (Alexander 2011a, 2011b; Bereiter & Scardamalia, 1993; Blakeslee, 1997; Brown et al., 2005; Dall'Alba & Sandberg, 2006; Geisler, 1994; Poe et al., 2010; Roth & Lee, 2002), beliefs (about writing, about the aims of science, and about the relationship between writing and science) may have a stronger influence on emerging scientists' capacity to gain disciplinary fluency than learned behaviours.

For many years now, I have observed a discrepancy in the way science writers are anecdotally represented as writers by those outside the scientific community, and the way they perceive themselves as writers. For example, Laura Martin, writing in *Scientific American*, comments, "professional scientists do not consider themselves writers. Some even brag about hating writing or being poor writers" (2012 para. 16). With a few exceptions, this has not been my experience within my professional life or in my research. Indeed, in an earlier Australasian study (Emerson, 2012), a majority (18 out of a sample of 20) of senior scientists reported very positive feelings about writing:

> I love writing. It's probably the part of the job that I love the most.
>
> I love to write—and to convey the passion I feel for my work.
>
> If I had the option, I would sit in my office all day and write.

Furthermore, of this group of 20 senior scientists (Emerson, 2012), 16 classified themselves as confident writers of science, who, even when they experienced the anxiety of writing for new audiences, nevertheless relished the challenge of doing so. Many of these senior scientists speak passionately and with insight about their work as writers of science and most of them see writing as integral to the construction of scientific knowledge. They hold complex views about audience, persuasion and the writing process.

My observation has been that some of our colleagues in the sciences are amongst the most sophisticated and flexible writers in the academy, often writing for a wider range of audiences (their immediate disciplinary peers, peers in adjacent fields, a broad scientific audience, industry, and a range of public audiences including social media) than most other faculty. And, since their writing is most often collaborative and multidisciplinary, their practices may be more socially complex, and require more articulation, mediation, and interpersonal communication, and more use of advanced media and technology than those of faculty in other disciplines.

And yet scientists' voices seem curiously absent when we talk about writing in science. With the exceptions of Burton and Morgan (2000), who conducted a survey and interviews with 70 mathematicians, and Yore and associates (2002, 2004, 2006; also Florence & Yore, 2004), who provide scientists' perspectives based on a relatively small sample, and a set of recent studies coming out of the literature on higher education related to graduate students' experience of writing a doctorate and co-authorship (e.g., Austin, 2009; Cuthbert & Spark, 2008; Golde, 2010; Kamler, 2008; Maher et al., 2008; Maher et al., 2013; Maher et al., 2014), research into scientists as writers focuses primarily on the observations and analyses of scholars outside STEM disciplines.

In my more pessimistic moments, I fear that our attitudes to scientists as writers have more closely resembled those of 19th century missionaries rather than anthropologists (Segal et al., 1998). In the context in which I write, Aotearoa New Zealand, the indigenous people, the *tangata whenua*,[2] are very cautious about *pakeha*[3] New Zealanders speaking about/for or interpreting Maori culture. The Maori people, like indigenous people elsewhere, see their culture as a *taonga,* a treasure, and only those in possession of this treasure, who own it as an integral part of themselves, are considered appropriate speakers for their culture. Anything else is cultural appropriation. Yet it has seemed to me, at times, that we in the humanities, who consider writing and rhetoric to be our own field, have no such qualms about the cultural appropriation of scientific writing and the experience of writing science.

This book is a partial attempt to address what we might see as the humanities' cultural appropriation of science writing. It invites scientists to speak of their beliefs, attitudes, processes, development and experiences as writers of science through the development of literacy narratives—and it invites those of us in the humanities to listen. By presenting science writers through oral history/literacy narrative, and as extended texts, rather than solely as interpreted by a writing researcher, this book provides a rich resource for scholars, teachers, and students, in both the sciences and the humanities.

ADDRESSING THE NEEDS OF UNDERGRADUATE STUDENTS

We need to hear these voices because the literacy aims of many of the undergraduate students in our writing classrooms include joining a scientific community of writers. Corkery (2005), following Shaughnessy (1979) and Bartholomae (1985), discusses student distrust and scepticism of the writing teacher's role of privileged gate keeper into the academy. But that scepticism may also arise—perhaps primarily arise—from a science student's perception that the writing teacher's knowledge of the genres, language, aims, and cultural practices of science, is limited. Why would any science student—some of whom may have a deep fear of or resistance to writing and who may fight away their fear by describing writing as irrelevant to a career in science (Poe et al., 2010)—trust a humanities professor who says "you can do this—and I can show you how"?

The Writing across the Curriculum (WAC) and Writing in the Disciplines (WID) movements have made considerable progress in repositioning the teaching of writing within the context of a specific discipline, giving cultural credibility and disciplinary relevance to the teaching of writing (Bazerman & Russell, 1994; Carter, 2007; Deane & O'Neill, 2011; McLeod & Soven, 2000). Yet still, what remain missing are texts written by scientists or mathematicians exploring their own pathways to competency in writing in their discipline. Corkery (2005) and Soliday (1994) argue that texts which present this kind of developmental narrative, such as literacy narratives, provide models of individuals transitioning into specific forms of literacy and thus provide motivation and models for students struggling to see their own transition into academic writing. This may be especially important to undergraduate science students, whose confidence in their own abilities as writers may have been damaged by experiences with writing in the classroom during their schooling (Choi et al., 2010; Shanahan, 2004). Several of the scientists and mathematicians in this study discuss damaging experiences with school and English teachers in particular. The anxious mathematics student, sitting in a writing class, who reads this comment by a successful applied mathematician,

> What's interesting is I did mathematics, I think, because I found English so difficult . . . I failed . . . on English and I was fine on mathematics. I was top in maths but I was desperate in English. I can remember the essay. The title was "Your House." Now as a mathematician . . . I've got to write about my house. What is my house? And I went to numbers straight away. It's got five windows, it's got one door—this is age 10 or 11. I knew it was a disaster when I wrote it. But I was incapable of doing anything better—Timothy, Chapter 3.

may recognise a similar incident of their own, and may never have realised that the successful science or mathematics professor in their writing classroom may have experienced this kind of setback. Reading of the way this senior professor worked his way to a position where he could ask for help (many times) and learn to write fluently in a range of genres, and finally say, "I'm pleased that I've got a job where I have to write," may provide the motivation to persist with writing in their discipline.

ADDRESSING THE NEEDS OF GRADUATE STUDENTS

For graduate students, who are beginning their apprenticeship within their disciplinary environment, these voices are also important. Research shows that novice science writers' induction into scientific writing is often poorly articulated, and dictated by the capacity or inclination of specific thesis advisors or peers to engage with writing (Florence & Yore, 2004; Kamler, 2008; Paretti & McNair, 2008) or the student's capacity for imitation (Alexander, 2005). Burton and Morgan (2000, p.450) comment that:

> Current practice in the training of mathematicians and in mathematics education more generally does not explicitly involve teaching and learning about mathematical writing. The novice may learn through using the existing models of published writing, through an apprenticeship of collaboration

with more experienced writers, or through the often harsh process of peer review. None of these methods is designed to help learners to acquire the kind of knowledge about language that might enable them to be aware of what they might achieve by choosing to write in different ways.

While graduate students might turn to resources on how to write scientific and mathematical documents (e.g., Penrose & Katz, 2010; Blum et al., 2006; and Day & Gastel, 2006), Morrs and Murray (2001) and Bishop and Ostrum (1997) both comment on the scarcity of research exploring the writing process of academics in general, suggesting that there is a gap between the writing processes described in such texts and "the real contexts and practices of [academic] writers" (Morrs & Murray 2001, p. 37). Similarly, Burton and Morgan (2000) show that specific directions relating to style provided by mathematical journal guidelines are seldom followed by mathematical writers who are published within those journals, suggesting such guidelines do not provide sufficient or appropriate direction for the novice writer.

Mutnick argues that literacy narratives are "a potential source of knowledge about realities that are frequently misrepresented, diluted, or altogether absent in mainstream depictions" (1998, p.85). Because narrators in a literacy narrative construct the past from the vantage point of present cultural knowledge and practice (in this instance, because the senior scientist constructs their narrative of their development as writer from a position of having developed rhetorical fluency), they are able, at the prompting of a skilled interlocutor, to reflect on, and draw attention to, the difficulties of making the transition to disciplinary literacy (Soliday, 1994), while presenting solutions that they know are successful. Thus they are invited to examine a process that is largely unarticulated within scientific and mathematical communities.

One final point in relation to graduate students: Corkery (2005), following Couture (1999), argues that the voices of disciplinary experts, portrayed through literacy narratives which show how someone, like the student, outside of academia brought themselves into it, "suggest different ways to bring the student's particular circumstances into an academic forum" (p. 57). Furthermore, as Couture argues, they provide a model or a vehicle for achieving community:

> Writers need to know . . . what it is that others *do* when they communicate in writing . . . and perhaps equally important, [so they can] *be* like them in order to occupy a common field within which each other's communications are heard and understood. (p. 42)

Voices portray more than strategy; they portray shifting beliefs, attitudes and character. A constructed character, yes—but nevertheless, a distinctive and individual character that conveys specific difficulties, struggles, triumphs and successes. One of the features of the narratives in this text is that many of the most successful writers (for example Elizabeth in Chapter 4 and James in Chapter 2) still struggle with writing issues and yet exhibit tenacity and determination (Daley, 1999; Florence & Yore, 2004), continually inventing and engaging new strategies to enable them to overcome these difficulties. A student confronted with a senior scientist sees only the successful present; a literacy narrative can reveal a more complex and conflicted image—an image which the student can meet on common ground. One of the most vivid examples of this is Lizzie, in Chapter 5. Her graduate students, struggling with their own writing, who are likely only to encounter her bright, capable and forceful professional presence, may learn more about how successful scientists also struggle with writing for a complex audience and be encouraged into resilience by hearing the story of how she spent a week in her pajamas wrestling a paper into shape.

Bereiter and Scardamalia (1993) argue that expertise emerges from learning communities that engage not just with the how, but the *why* questions in the field of expertise. Literacy narratives invite scientists to engage with the *why* questions of writing (for example, linking distinctive disciplinary genres to the aims of science), providing a starting point and a language for research communities to engage with the deeper questions relating to writing in their discipline.

For both undergraduate and graduate science students, then, hearing the voices of scientists is vital, in terms of motivation and resilience, providing models, and becoming part of—and developing—a scientific community that has a language to speak about writing.

DISCIPLINARY DIFFERENCES RELATING TO PROCESS

For teachers in composition and WID, these voices may teach us something new about disciplinary *process*. While writing teachers are acutely aware of different disciplinary genres, there is a danger of assuming that process is process; i.e., that while genre, style, structure, and audience may differ across disciplines, process is transferable. It's possible that this is incorrect (Driscoll, 2011; Melzer, 2014). Several of the participants in this text distinctly highlight the differences both in context and process between humanities-based writing processes and scientific writing processes, largely around the issue of collaboration and co-authorship (Austin & McDaniels, 2006; Jacoby & Gonzales, 1991; Maher et al., 2013; Maher et al., 2014). Kamler (2008, p. 288), in a study which compares the publication processes of graduates in different disciplines, comments: "For [the] science students, co-authorship was a given . . . a crucial part of learning the ropes of academic publishing . . . [a] well-established expectation of the discourse community." And while collaboration, conversation and peer review are very much part of the language of composition, the context in which they take place in the sciences (co-authorship, the hierarchies of disciplinary or interdisciplinary teams, the drafting process, and the use of technology such as the use of LaTeX) are rarely discussed. There are few parallels in the humanities for the following activities:

> Much of my writing now is first drafted by or with someone else: all of the team will create the story. Someone has to start with a draft. What I will do, particularly with my graduate students . . . [is ask] "now, look, what is the story? What are the pictures and so on?" . . . So . . . I'll say . . . "we agree on the story" (that's a discussion, right . . .) but when the writing actually starts I'll say "look, here's an introduction. I want you now to go away and write the rest of the paper." So, they will start, and the next thing will tend to be what was the experimental method?, what were the results? and so forth. Then we'll start getting the more difficult stuff about the interpretation of that story, and how we would end it off. In the process, it will go backwards and forwards. We don't sit down and write together. I write something, they add something on, I will correct that or make suggestions, sit and talk with them, they'll have another go. We'll go backwards and forwards—
> Richard, Chapter 2.

In our search for common ground between writing and STEM faculty (Poe et al., 2010), and in the relief of finding common terminology, such differences in process may be overlooked. For writing teachers, these voices provide us with new perspectives of what writing in the sciences involves, and how process works in a scientific context, and perhaps suggest new approaches to curricula that will meet the needs of our students more effectively.

THE GOALS AND METHODOLOGICAL BASIS OF THIS BOOK

> The goal and responsibility is to evoke and bear witness to a situation the researcher has been in or studied, inviting the reader into a relationship, enticing people to think and feel with the story being told as opposed to thinking about it.
> — Smith & Sparkes, 2006, p. 185

The principal aim of this book, then, is to break down the science-humanities and senior researcher-student divides in a way that speaks to the idea that we are all writers and that we are all engaged in a journey to rhetorical fluency within our disciplines. By conveying, in their own voice, scientists' stories as writers—their beliefs, attitudes, and experiences—my intention is to invite readers "into a relationship" with the storytellers. I'm primarily concerned, not just with the generalizable, but with the particular experience. Hence, the voices in this book are not intended as a representative sample, but as indicative of the wide range of university scientists' voices, in terms of discipline, experience, seniority, institutional type, and geographic location.

Method follows aims, and so the following sections outline how participants were identified, how the stories were collected and managed, and why the stories are conveyed as the hybrid oral history/literacy narrative genre.

THE PEOPLE IN THIS BOOK

The narratives presented in Chapters 2–6 of this book are developed from a selection of interviews conducted with university scientists between 2009 and 2014. In all, I conducted 106 interviews with scientists in a range of different scientific disciplines from 17 universities in North America, the UK, and Australasia. These 106 comprised 62 senior scientists and 44 junior scientists (made up of 26 emerging scientists no more than seven years out of their doctorate and 18 Ph.D. students in science). The scientists in this sample came from a wide range of disciplines, but a decision was made to exclude pure mathematicians in the sample, since preliminary analysis of early interviews suggested that their experiences of writing were so distinctive that they would need to be explored in a separate study.

In terms of demographics, 37% of the sample (a total of 39) were women (16 senior scientists, 13 junior faculty/postdocs, and 11 Ph.D.s)—reflecting the demographic trends of women in science (Bentley & Adamson, 2003; Hill et al., 2010; Maher et al., 2014). Of the sample, 35% was working in North America at the time of interviewing, 41% in Australasia, and 23% in the UK, although these figures do not fully represent the diversity of the sample: as we might expect of an international community, many were working and/or had studied outside their country of origin. The majority had English as a first language—only 10 of the participants identified English as a second language and, of this group, only two (one from China, one from Argentina) were not fully fluent. The choice of primarily participants who had English as a first language was intentional: English is the language of science (Gordin, 2015; Montgomery, 2013), which means that non-English speaking scientists must learn to write science in English, but exploring the difficulties inherent in scientists acquiring or refining a second language while learning to write science would be a major study in its own right and is beyond the scope of this text.

The process of choosing participants for this research was multifaceted and based on the concept of purposeful sampling (Leydens, 2008). In each institution I identified the senior scientists first. In some institutions, my contacts in the writing program made recommendations or arranged interviews. Beyond these recommendations, I looked for participants from three groups within a range of science disciplines: those with an extensive publication record in their field, those who showed a particular interest in cross-disciplinary writing or writing for a public audience, and those with a moderate publication record. There was a snowball effect: some participants recommended other people they thought I should interview; other participants heard of my research and contacted me to volunteer. My experience was that if I found the senior scientists in an institution to engage with this study, they would help me to locate the emerging scientists or the doctoral students in their discipline. The sample, therefore, was not random, and was, to some extent based on self-selection. It is important to say again that the scientists whose voices appear in this book are not a representative sample of the scientific community—but then I never set out to achieve that. Instead, what I have collected here are a range of voices that are intended to show a wide variety of perspectives within the context of the university-based scientific community.

COLLECTING THE STORIES

Information was gathered from each participant using two methods. Once a participant had agreed to take part in the study, a short survey (Appendix A) was sent by email to the participant to collect demographic data and task-related information. Following this, a time was scheduled for a semi-structured interview which would form the basis for the participant's narrative: this was the central data collection method.

A number of different methodologies could have been used to fulfil my aim of allowing scientists to speak into the literature about science and writing. Attitudes and beliefs can be collected through quantitative and qualitative survey data. Even experiences and writing development themes could have been collected through written data collected in response to directive questions.

Nevertheless, such approaches would provide only a fragmented and limited picture of participants' perceived experiences of scientists as writers (O'Shaughnessy et al., 2012). Anxieties and doubts are unlikely to be explored using such methods, and participants may provide what they view to be the "correct" or socially acceptable answer to questions, rather than relating their lived experience and perceptions, without the intervention of follow-up or probing questions. Given that writing was likely to be a difficult or contested topic for some participants, a method that allowed them to gradually explore their experience, prompted by appropriate questions that emerged from their discussion, was likely to produce richer and more valid data. The choice to use semi-structured narrative interviews was based on a desire to elicit rich, complex data about scientists' perceptions of their own experience as writers of science.

A further issue relating to data collection was motivated by a desire to focus on what is often a forgotten backstory of scientists' professional lives. O'Shaughnessy et al. (2012, p.44) observe that participants may have both dominant and subjugated stories about their lives, and that narrative interviews are an effective way "to make more conscious aspects of their lives which have been dismissed as not important." My participants' dominant professional story relates to their primary identity as scientists. Focusing on a subjugated story, their story as writers of science, was the central focus of this study. Scientists may not always identify as writers, and some scientists may not consciously recognise that their professional lives are constructed on written texts; the interviews in this study showed that most had never discussed their writing practices and experiences. A narrative interview allows the participant to become more self-aware during the interview process, to see aspects of themselves and their own practice to which they had not previously paid attention (Bleakley, 2005; Rich & Grey, 2003). This was undoubtedly the case for some of the participants in this study. Several reflected on the unexpectedness of their own answers. One participant commented, "you are doing a very hard thing. You are asking us to be self-reflexive about things we don't normally think about. I'm glad to have talked to you today." By having someone ask questions about their writing, listen attentively to their responses, and ask challenging questions, some participants were able to discover things they didn't know they knew or thought.

Sometimes this thinking went on well after the interview. Participants sent me follow-up emails, clarifying or changing the points they had made after further reflection. One emailed me several weeks after the interview, to explain how she'd been thinking about the interview and how it was changing her perceptions of her own work:

> I've been intending to follow up on our meeting for weeks now. I just wanted to say what a profound conversation it was for me. I had several interesting "aha!" moments in talking with you that I've shared with others since then. One was in acknowledging that I enjoy writing but, having said that, how

difficult I find it to be. . . . The second was in articulating how much writing I do for others . . . letters for people, reviews of papers and grant applications, editing my students' work, etc., and the recognition that it's as legitimate a part of my work as writing a first-authored paper (even if I don't believe that in my heart of hearts yet). (Senior Scientist, Conservation Ecology)

Others sought me out to do a follow-up interview (sometimes after a discussion with others) or to tell me how our discussion had affected their teaching or graduate supervision due to an increased awareness of the relationship between writing and science which had emerged through their own thinking. Data from these follow-up conversations were also collected and recorded if appropriate.

The interviews were semi-structured around themes of writing process, attitudes and beliefs about writing and science, and learning to write science (Appendix B and C). Interviews were scheduled for one hour, but lasted between 40 minutes and 3 hours (the latter conducted over three sessions). Most interviews were conducted face-to-face, but three were conducted by telephone and one by Skype. The interview question sheet was continually adapted as new issues arose with the participants. New questions were added, such as a question about the writing lifecycle, after it emerged from one early interview, while other questions were dropped if the answers to them seemed unreliable or if they consistently failed to contribute to the overall story. Where possible, I encouraged participants to tell specific and detailed stories about their experience of writing in science. I researched each individual prior to the interview, read some of their work where possible, and included specific questions about each person's writing history.

Participants responded differently to the interview process. Some participants gave concise answers to each question, some were ready to speculate and self-reflect at considerable length, while others resisted the interview format and/or the questions and took control of the interview from the start. Richard, for example, whose narrative is provided in Chapter 2, blithely batted aside many of my questions and discussed issues that he considered important that were not covered by the interview plan. In instances like this, I felt it important to respect those choices, and to engage spontaneously with the participant's narrative, adding in my questions, where possible, in whatever order seemed most relevant to the discussion. If some questions therefore weren't addressed, I did not push the issue. Mason (whose narrative appears in Chapter 3) objected to most of the questions I asked, on the grounds that they made assumptions that were not consistent with his experience and beliefs. Again, it was important to respect those choices and reflect on his objections, but also to probe the resistance in an effort to gain more understanding of the perceived disjunction.

One difficulty which emerged was that many participants volunteered information or asked questions after the recording device was switched off. In this situation, I would write notes on the additional discussion after I left the office, add these to the transcript and, if the interview was to be revised into a narrative, ask participants to check my interpretation and documentation of the discussion and approve the addition.

MANAGING THE DATA

The data were gathered and then initially processed using a two-stage process: transcript correction and analysis, and (for some interviews) construction into narratives. All the interviews were transcribed and checked, adjusted for anonymity, and then edited for comprehension. Digressions (for example, long discussions of their field, unrelated to writing and science) were removed or compressed, and the transcripts were then revised.

Of the 106 participants in this study, only 19 full or extended narratives are included in this text. The interviews selected for development into narratives were chosen primarily to represent a range of themes while giving some consideration to geographical and disciplinary breadth. The procedure for narrative construction was that most of the interview questions were removed and integrated into the narrative as needed for sense, and repetitious phrases which didn't add to the meaning (e.g., repetitious use of "I guess," "sort of," "kind of") were reduced. However, since my aim was to retain the participant's "voice," some phrases used regularly by the interviewee were retained in the narrative.

The finalising of the narratives was an iterative process. It was important that the participants felt ownership of their narrative. Narratives were sent to the participant for checking and approval. Track changes were used to direct the participant's attention to areas that needed clarification or to acronyms or names that would not be accessible to the general reader. Once the participant had returned the narrative, their changes were incorporated into the text and sent back again for re-checking. Occasionally this happened several times. When the narrative had been finally approved, I added a theme, taken precisely from the narrative, and an introduction to the speaker.

Finally, I asked each narrator to provide a pseudonym. Some were happy for me to choose a random name; others chose names that were in some way significant to them, which is why some narrators have one and others two names. Choosing a pseudonym was an important process for some of the narrators, leading to much discussion; some chose names related to a childhood memory, names given to them by significant (personal or professional) others, or names that represented someone who was significant to them.

One narrative needed special care. Richard died between the interview and the completion of the narrative. I contacted Richard's family and asked for permission to use the narrative. I invited them to make any adjustments they felt were appropriate (they made none), choose a pseudonym, and recommend another scientist to check the narrative for technical correctness. They suggested that Cameron, who worked closely with Richard, be invited to take on this role. Both the family member I worked with and Cameron remarked that it was a bittersweet moment to hear Richard's voice speak so distinctively through this narrative.

PRESENTING THE NARRATIVES

> Stories are knowledge.
> — Moana Jackson[4]

In determining how these voices were to be portrayed in the text, the options I considered were *story analysis* and *storytelling* (Smith & Sparkes, 2006). Story analysis is the more conventional approach, leading to a scholarly text based on thematic analysis. In a scholarly analysis of narratives, the researchers step "outside or back from the story, employ analytical procedures, strategies and techniques in order to explore certain features of the story" (Smith & Sparkes, 2006, p. 185). Story analysts write *about* the stories (Frank, 1995) and approach the narratives from a methodological standpoint, seeing narratives as reliable (or unreliable) forms of data (Atkinson, 1997; Smith & Sparkes, 2008). In story analysis, the primary voice is the researcher's, and the researcher controls the narrative and analysis, moving from the particular to the generalised.

Such an approach proved problematic for me, both in terms of my own ethical position as a researcher and the aim of the text as providing a bridge between disciplines and between writers. I felt a text based primarily on the product and process of a story analysis to be a form of cultural appropriation. Furthermore, a story analysis would limit the immediacy of the narratives and the readers' capacity to engage, rationally and empathically, with the narrators' lived experience as writers. While story analysis is included in this text (in the introductions to Chapters 2–6, and the final chapter, which takes a broader perspective on the data set as a whole), another approach to working with the data was needed to achieve my goals.

Storytelling, by contrast, sees the story as the product of the enquiry, and invites the reader *into* the story, to engage at both emotional and rational levels with the narrator's experience (Frank, 2000). The researcher-as-storyteller understands the story itself as containing analytical techniques, theory, and dialogical structures (Bleakley, 2005; Ellis, 2004) which can speak for themselves:

> In a narrative analysis, storytellers emphasize that participants' stories of the self are told for the sake of others just as much as for themselves. Hence, the ethical and heartfelt claim is for a dialogic relationship with a listener (including the researcher) that requires engagement from within, not analysis from outside, the story and narrative identity. Consequently, the goal and responsibility is to evoke and bear witness to a situation the researcher has been in or studied, inviting the reader into a relationship, enticing people to think and feel *with* the story being told as opposed to thinking *about* it. (Smith & Sparkes, 2006, p. 185)

The decision to include extended narratives as the primary structure of this text was motivated by these attributes of a storytelling approach to narrative: the invitation of the readers into the story, the possibilities provided by a dialogic relationship in evoking changed attitudes in the reader, the goal *and responsibility* to "bear witness" to these conversations (Bleakley, 2005). This is the primary motivation of this book: to invite the reader (whether that is a teacher/writer from the humanities, an undergraduate or graduate student of science, or a senior scientist) into the story of scientists' experiences as writers of science.

Perhaps the key feature of storytelling for this study is its view of the construction of the narratives as dialogic (Stephens & Breheny, 2013). Interviewing is not a "transparent process of information gathering" (Bleakley, 2005) and it is important to acknowledge the impact of the researcher on the construction of the stories (Mishler, 1995). The narratives in this text were co-constructed by myself and the participants. I provided both the opportunity to dialogue and the initial questions, responded spontaneously to their answers, occasionally engaged in argument, and eventually shaped the narratives through the editing process. Despite the iterative consultation, my hand is on the narratives at every step, and it is important not to minimise my contribution to the final story. My own story, as an outsider (I am not a scientist) means that participants may have simplified their process for the sake of a non-scientist.

Nevertheless, the role I took, of a naïve interviewer, as someone outside of the experience of the participants, is appropriate to the readership of this book, which includes writers and writing teachers from other disciplines, and emerging scientists, who also stand on the edge of the experiences of the narrators in this text. This issue of the role of the interviewer is explored further in the following section.

AN IMPORTANT ISSUE: RELIABILITY

Because the interviews investigated a subjugated and emerging story, some of the narratives in this text contain internal contradictions, even among participants who appeared to have the clearest and most well-rehearsed story about writing. Examples of this can be found in Richard's discussion in Chapter 2 of whether a bad writer can be a good scientist, and Timothy's portrayal of himself as both a confident and struggling writer in Chapter 3. I have made no attempt to remove these contradictions. While cohesion is often seen as an essential form of validation in narrative theory (Harvey et al., 2000; Holstein and Gubrium, 2000), I was aware that participants were telling a new story, struggling to find coherence about a topic they rarely thought about, and that this meant that emerging thinking could lead to apparent lack of cohesion (Baerger & McAdams, 1999). Sometimes my questions broke through a carefully constructed narrative; Timothy, again, indicates that he hadn't expected to talk about his experiences of writing in childhood and these questions led to the emergence of a new story. In most instances (but not all) the participant became aware of the contradiction within their narratives and attempted to achieve cohesion.

Nevertheless, the issue of reliability is important in considering the nature of these narratives. Clearly there was no need to objectively verify described events, since our focus is on the subjective experience of participants and the meaning they attributed to those experiences. For example, Jane, in Chapter 3, describes her graduate advisor as hostile and unhelpful during an oral examination, but no external verification of her experience is provided. It is possible that others at this meeting may have experienced his behaviour differently (although she does provide some evidence of other people's experience of his behaviour in other contexts), but our interest is in how Jane experienced her advisor's behaviour and the way she portrays its impact on her confidence as a writer.

One of the challenges of the interviewing process was to somehow break through real barriers to describing personal experience: people forget or minimise painful past experiences in the more confident present, and professional personas are valuable public shields against personal insecurities or uncertainty. I knew that, for some participants, writing would be an area of personal and professional insecurity (Florence & Yore, 2004; Yore et al., 2006). Being aware, as an interviewer, of strategies to avoid personal disclosure was essential throughout each interview. For example, a common strategy used by the participants to avoid personal engagement was presenting a generic or "correct" approach to writing rather than recounting their actual experience or strategies. When asked about writing a scientific paper, some participants would speak in the second person ("well, you write the method, and then then you look at the figures") or use indefinite pronouns ("mostly one starts with the pictures"). The use of a second person or indefinite pronoun usually (but not always—some participants spoke predominantly in the second person) alerted me to the possibility that the participant was not discussing their own process but rather the process they believed to be correct. Such a response also often lacked the detail I was looking for. To counter this, I did not challenge their approach but later in the interview asked participants to think about a recent writing project and then to walk me through each step of the writing process—or I asked them to tell me a story about their first paper or their most challenging writing task. The latter was especially effective for talking to people who insisted that learning to write science was "effortless"—more often than not, a question about a specific writing experience elicited a quite different response. It was also sometimes necessary to ask detailed questions to bring participants back to their actual lived process, to ensure that the details of the process were captured.

While it is important to see the reliability of the stories in relation to the subjective experience of the speaker, and in terms of an emergent, subjugated story, internal contradictions in some narratives invite the reader to question the reliability of some central ideas within the narrative. Jane's story, for example, contains a central theme that she is alone, unsupported, and courageous in her attempts to find ways forward as a scientist and as a writer. However, there are elements in the story that bring this theme into question. Various people offer her support—an editor provides meaningful feedback, a colleague sets up a writing group, a scientist from another country with whom she has no previous professional relationship volunteers to visit her and works with her over several years, two colleagues help her write a significant grant proposal—which suggests she is not entirely alone and unsupported. Some of her comments ("I have a fear of the blank page and I'll do anything to procrastinate my way out") suggest that her limited publication record is caused by internal rather than external factors, while her discussion in the penultimate paragraph of her chair's proposed initiative to get groups and peer support strategies in place in the department suggests a lack of agency on her part. This does not suggest, of course, that Jane does not experience herself as alone, unsupported, and courageous, nor that she has been denied supports (e.g., the support of a helpful advisor and an effective research team) that she was entirely justified in expecting within the cultural context of her professional life. But it does suggest that there may be elements in the story that the narrator is, to some extent, unaware of, and that we should read the stories carefully.

One final issue concerning reliability: in any interview situation, the interlocutor is in some sense constructed by the participant, and this may affect the participant's responses to questions. In this case, the participants may have at times provided answers in terms of their expectations or assumptions of the interviewer based on such issues as gender, nationality, or disciplinary-allegiance. The latter was most clearly evident in one particular question about reading for pleasure: those who did not read outside of a professional context, or those who read only news and non-fiction, were sometimes reluctant to admit this, or apologetic about what they feared I might perceive as a personal shortcoming. On the other hand, this also opened up possibilities. For example, one participant told me about a piece of creative nonfiction he had been working on for years, and commented with pleasure, "I've never told anyone about this before!" While the impact of the constructed interlocutor can never be eliminated, or even entirely measured, and may be different for each participant, my aim was to offer a non-judgmental, interested, curious, and naïve presence, and to be alert to the signs of its impact (positive and negative) on the stories being constructed.

In some essential ways, then, such as the narrative technique and the constructed interlocutor, these stories are dramatic monologues. As such, they invite the reader into an experience that is both empathetic and critical.

Chapters 2–6 of this study are focused on storytelling. Each chapter presents 3–5 participant narratives based around central themes: public-focused writing, the reluctant writer, the writing community, the development of the scientific writer, and creative writing. The aim of these chapters is to invite the reader into the scientists' experience of writing and learning to write within a disciplinary context. While each chapter gathers narratives with a particular focus, I have not edited the narratives solely to illustrate that theme. In Chapter 2, for example, while the narratives have been chosen to illustrate the theme of public engagement/writing, the narrators also discuss more fully their experiences of writing and learning to write. My intention here is to convey a scientist's interest in public writing *within the broader context of a scientist's perspective on writing.* This approach is central to a storytelling methodology: a scientist's choice to engage with public audiences emerges out of—and needs to be seen within— their beliefs, attitudes and experiences of writing, the way they learnt and were mentored into writing. For this reason, we need to hear not just disembodied segments of narration but a contexualised theme embedded in a full and rich discussion of the scientist's experience.

READING THE THEMES

Beyond this interest in the particular lived experience, however, there is also obvious value in considering the themes that emerged from such a substantial data set, and these themes are addressed in Chapter 7, using a story-analysis approach. Using themes that emerged from the literature and from a preliminary

Chapter 1

Quadrant 1: Early influences	Quadrant 3: Attitudes
Childhood attitudes/experiences of writing	Enjoyment
Undergraduate attitudes/experiences of writing	Motivation
	Resilience
	Self-efficacy/purpose
Quadrant 2: Learning to write science	**Quadrant 4: Beliefs**
Advisor	Function of writing
Community	Audience
Rhetorical reading	Persuasion
Ongoing support post-Ph.D.	Beliefs about identity/role as a scientist.

Figure 1.1: Model of a scientific writer

analysis of the data, the analysis of the entire data set is based on the following model (Figure 1.1).

The remainder of this chapter focuses on these central themes and, in particular, research to date relating to each theme, and the gaps that this study will address.

EARLY ATTITUDES

Much of the current research on scientists' perceptions of themselves as writers is narrowly focused in three areas: experiences of undergraduates as disciplinary learners, primarily through the literature on WAC and WID; experiences of doctoral students, through the literature on graduate supervision; and research which examines the doctoral/post-doctoral (or novice/expert) divide. A consequence of this narrow focus is that the picture that has been drawn of scientists' development as writers is a relatively simple one: scientists may begin to learn to write science as undergraduates through writing-intensive classes, but the graduate years are a central place in which scientists learn the craft of disciplinary writing, through a process of cognitive apprenticeship and mentorship (Burton & Morgan, 2000; Cummings, 2009; Cuthbert & Spark, 2008; Florence & Yore, 2004; Maher et al., 2013; Poe et al., 2010).

We will return to the issue of undergraduate/graduate experiences later in this chapter, but pause here for a moment to observe one limitation of the current narrow research focus on scientists' perceptions of themselves as writers:

what of scientists' earlier experiences? How do they impact scientists' development as writers?

In the preface to this study, I noted that my undergraduates are surprised—and perhaps somewhat dismayed—to find writing is part of their science curriculum. Similarly, Poe at al. (2010, p.1) begin their study by commenting "MIT students, by and large, do not love to write . . . the science and engineering orientation of MIT undergraduates can often lead them to believe that in their professional careers, the search for engineering solutions or scientific phenomena, not the seemingly tedious process of communicating those findings, will dominate." If such observations are correct, these attitudes and beliefs—including a reluctance to write, and a belief that writing is somehow separate to and unnecessary for science—come from somewhere.

And yet, the early educational experiences of scientists in relation to writing appear to be largely unexamined. Instead, as with the two examples cited above, the evidence seems to be largely anecdotal. Laura Martin (2012, para. 9), for example, suggests that the division between science and writing begins at elementary school, and that, consequently, children with an interest in science are separated from writing-rich subjects at an early age:

> On top of rigid conventions, scientists must contend with the pervasive myth that scientists can't write. We begin differentiating scientists and writers in elementary school. One "likes math" or "likes English." Our academic system, from pre-K through graduate school, contrasts science and literature—objectivism and subjectivism, reductionism and holism.

Martin provides no empirical evidence for such a view, and in some countries there is evidence that contradicts such a perspective. In New Zealand, for example, the national curriculum embeds writing and communication into every discipline at both elementary and secondary school (although how that works in practice may be another matter), and the pedagogical value of integrating writing for learning in science in schools has been extensively researched (see, for example, Bressler, 2014; Chinn & Hilgers, 2000; Choi et al., 2010; Holbrook & Rannikmae, 2007; Prain & Hand, 1999; Shanahan, 2004; Rowell, 1997). Nevertheless, anecdotal evidence suggests that, somewhere in their education, students with a science orientation adopt—or may be encouraged to adopt—particular attitudes and beliefs about the nature and purpose of writing and its relationship to science which need to be challenged if they are to become proficient writers of science (Shanahan, 2004). Beliefs and attitudes inform praxis and are resistant to change (Brady & Winn, 2014; Martinez et al., 2001; Tobin & Tippens, 1996), as we discuss more fully later in this chapter.

Beyond the experience and impact of schooling, the undergraduate years have recently been seen as a significant opportunity to embed writing into the curriculum, and writing or communication-intensive courses as they pertain to the sciences have been extensively researched, including students' experiences of those courses (Bayer & Curto, 2012; Bayer et al., 2005; Hand et al., 2001; Reed et al., 2014; Reynolds et al., 2009; Reynolds et al., 2012; Stanford, 2013). However, the long-term impact of these programmes on scientists' attitudes to and beliefs about writing, and their impact on praxis, remains largely unexamined. The extent to which these learning initiatives are reaching the majority of undergraduate science students is also something of an unknown (Thaiss, 2012).

By taking a broader approach to the question of influences on scientists' development as writers, this study intends to examine these early years: to what extent did early learning experiences and early attitudes to writing, formed during the K-12 years, impact on scientists' later development as disciplinary writers? And how do scientists perceive the impact of their undergraduate education on their development as writers?

LEARNING TO WRITE SCIENCE

Learning to write within a disciplinary academic context is part of the process of being socialised into a disciplinary community (Bazerman, 1992; Kelly, 2007; Lee & Aitchison, 2009; Maher et al., 2013; Maher et al., 2014; Norris & Phillips, 2003). Austin and McDaniels (2006, p. 400) describe this socialisation process broadly as "internalising the expectations, standards, and norms of a given society," including relevant skills, knowledge and behaviours, attitudes, beliefs, and values. In examining how scientists become writers, therefore, we must consider both the individual and the social context, especially given the collaborative nature of much scientific writing (Cummings 2009; Golde, 2010; Maher et al., 2014).

Bereiter and Scardamalia (1993) suggest a nurturing and challenging learning community, or "community of practice" (Dunbar, 1996; Riel, 2000) is essential if writers are to develop the flexible writing skills required in the sciences (Hatano & Inagaki, 1986; Holyoak, 1991). The ability of the scientific writer to demonstrate an interest and capacity to engage with a range of discourse communities in a range of genres should, they suggest, emerge from a deeper conceptual understanding of the domain borne out of a community of practice which favours deep engagement and learning. Such a conceptual understanding is effected by a learning context or community in which understanding is favoured over goal completion, where tasks are variable and to some extent unpredictable, and where there is social support which encourages exploration, experimentation, and deeper comprehension through discussion of *why* questions rather than simple *how* questions. Poe et al. (2010, p.10), following Riel (2000) and Smith et al. (2005) build on this concept of the learning community, describing it as:

> A setting in which the community is organized rather than disciplined; characterized by collaboration rather than competition; focused on knowledge construction rather than knowledge delivery from one central source; student centered rather than teacher centered; interdependent rather than independent. Instead of expertise flowing from the teacher to many students, expertise flows in many directions . . . leaders are people who inspire others to work towards common goals.

Broadly speaking, the research to date suggests that the cultural context of science is not wholly conducive to the development of its emerging members as scientific writers. Maher et al. (2013) comment that enculturation into any disciplinary discourse practices is no simple matter, but that there may be additional challenges in the sciences related to social context and cultural and pedagogic expectations (Collins et al., 1989). While the cognitive apprenticeship model, embedded within a learning community (Austin, 2009; Collins et al., 1989; Poe et al., 2010) is central to the doctoral process, its success as a model of teaching writing is open to debate. Within this model, reading and imitation (Burton & Morgan, 2000) and feedback from doctoral advisors, lab partners (Florence & Yore, 2004; Kamler, 2008), collaborators and peer review (Austin 2002; Burton and Morgan, 2000; Gardner, 2009; Sweitzer, 2009) are the primary ways in which students learn to write, with faculty-student co-authorship (Kamler, 2008; Maher et al., 2013) the "signature pedagogy" (Maher et al., 2013, p.128) of learning to write in the sciences.

While such approaches to learning disciplinary writing clearly can be successful (see, for example, the discussions of co-authoring in Florence & Yore 2004; Kamler, 2008; Maher et al., 2013), they may also be fraught with difficulty (Lee & Aitchison, 2009; Paré, 2011; Starke-Meyerring, 2011). There is evidence, as we have already observed, to suggest that some resources available to support emerging scientists as writers, such as style guides and journal guidelines, do not provide accurate or appropriate direction (Bishop & Ostrum, 1997; Burton & Morgan, 2000; Morss & Murray, 2001). Lerner (2007, p. 215) comments on one particular example (Knisely, 2005) where the author is explaining how the structure of a scientific paper facilitates reading:

> Although scientists' act of reading research articles might resemble this process to some degree, the emphasis on reading

as information gathering, and thus writing as information conveyance, captures none of the dynamism of the experimental process and largely ignores the complexity of the rhetorical choices open to a scientific writer.

Further, the process of writing collaboratively with a mentor is socially complex, incorporating hierarchical structures (Jacoby & Gonzales, 1991) that may or may not be open to revision (Florence & Yore, 2004. Co-authors may bring (sometimes unarticulated or even subconscious) expectations to the collaborative process; Maher et al. (2013), for example, list a series of advisor and co-author expectations related to writing skill level, prior experience of writing and reading, and beliefs about the purpose and process of writing and co-writing, which senior scientists describe as commonly not shared by the emerging scientists with whom they are co-authoring. Learning by imitation (e.g., by reading in the discipline) may be a hit-and miss affair. As Collins et al., (1989, p.7) observe, "students are unable to make use of potential models of good writing acquired through reading because they have no understanding of the strategies and processes required to produce such a text." Without access to these cognitive and metacognitive strategies, students must learn intuitively, and thus lack a capacity to articulate their rhetorical choices.

The current context of science, then, appears to offer limited, and largely unarticulated and unsystematised, support for emerging scientists as writers, with none of the methods currently available "designed to help learners to acquire the kind of knowledge about language that might enable them to be aware of what they might achieve by choosing to write in different ways" (Burton & Morgan, 2000, p. 450). While science is formed within the investigative learning communities that Bereiter and Scardamalia (1993) identify as essential to the formation of new knowledge, there is little evidence to suggest that these communities reliably work to articulate and support the advanced, flexible writing expertise that emerging scientists need. For some researchers, the co-authorship relationship is a potential strength that needs reframing:

> Co-authorship with supervisors is a significant pedagogical practice that can enhance the robustness and know-how of emergent scholars as well as their publication output. There is a need, however, to rethink co-authorship explicitly as a pedagogical practice. (Kamler, 2008, p. 283)

Others have suggested the need for a new approach, querying whether faculty mentors are the best people to support emerging writers (Maher et al., 2008), and whether a workshop (Maher et al., 2013), or writing groups model

(Cuthbert & Spark, 2008; Kamler, 2008; Maher et al., 2008) would be more effective ways of enculturating emerging scientists into disciplinary writing.

All this is to assume that the primary—or perhaps only—context in which scientists learn to write is during the years prior to the completion of a doctorate. It is possible, however, that the process of learning to write in the sciences has a longer duration, and that different factors may impact on scientists' development as writers post-Ph.D. Since it is unlikely that scientists could be considered fully fledged disciplinary writers when they hand in the dissertation, how do they continue to learn and develop as writers within the cultural context of science? For those who move into cross-disciplinary writing or writing for public audiences, and thus into writing in new genres, what support is available and how is that support accessed?

This study, then, investigates scientists' perceptions of their own development as writers within the cultural context of science. To what extent do they perceive the different aspects of the cognitive apprenticeship (reading and imitation, mentorship and co-authorship) to be successful? What do they perceive to have been the key influences on their development as writers of science? Is the post-Ph.D. period important to their development as writers—and, if so, what forms of support are available?

ATTITUDES

As stated earlier in this chapter, learning to become an effective disciplinary writer requires more than the acquisition of a set of skills and knowledge (Blakeslee, 1997; Dall'Alba & Sandberg, 2006; Gee, 2005; Poe et al., 2010); writing takes place within a social and cultural context (Collins et al., 1989; Doody, 2015; Maher, 2008) and in the context of forming disciplinary identity (Brown et al., 2005; Gee, 2005). For this reason, investigating the attitudes and beliefs of scientists and emerging scientists is vitally important (Kagan, 1990).

While, as we have discussed earlier in this chapter, scientists are sometimes anecdotally portrayed as reluctant writers, the broader limited literature on scientists' perceptions of writing calls this into question. Florence and Yore (2004), following Daley (1999) suggest scientists are driven, passionate about their field, continually dissatisfied with current understandings (Bereiter & Scardamalia, 1987, 1993), and therefore driven to write by their desire to contribute to a continuing disciplinary debate. Hartley and Branthwaite (1989) in a study of academic psychologists noted that the most productive writers had positive attitudes to academic writing and felt that their writing was important to them (see also, Hartley & Knapper, 1984). They noted that writing anxiety decreased with experience and productivity, and individuals who enjoyed writing were less anxious and most productive. Within the broader literature on academics' writing, Rodgers and Rodgers (1999) suggest that prolific academic writers are likely to enjoy writing, be energised by writing, and respond constructively to reviewer criticism. A sense of personal accomplishment and dedication (Fox & Faver, 1985; Jones & Preusz, 1993), resilience (Boice, 1994), confidence (Morrs & Murray, 2001; Shah et al., 2009), and intellectual curiosity (Veronikas & Shaughnessy, 2005) have also been identified as key characteristics of successful academic writers.

In the broader literature on expertise, attitudes are described as critical in the shift towards competency in any discipline (Alexander, 2011a, 2011b; Brady & Winn, 2014; Dreyfus & Dreyfus, 2004, 2005), with Dreyfus and Dreyfus (2005) and Benner (1984, 2004) describing emotional engagement as the starting point in the move towards expertise. Alexander argues that emotional engagement is as important as cognitive capacity in the development of academic expertise, with both competent and expert practitioners showing deeper affective engagement and more investment in their work than those who are at earlier stages of development within a discipline. Bereiter and Scardamalia (1993) suggest that experts are more curious, more likely to pose new questions, more robust and more tenacious and determined than experienced non-experts.

Alexander (2003, p.10), in her Model of Domain Learning, sees motivation as one of the central attitudinal features of academic expertise:

> individuals' motivation and affect are significant contributors to the development of expertise . . . without understanding those motivational/affective dimensions, educators cannot explain why some individuals persist in their journey to expertise, while others yield to unavoidable pressures.

She notes that motivation and interest have been largely overlooked in studies which have concentrated on the cognitive abilities, behaviours and skills sets of experts (such as those by Dreyfus and Dreyfus, 2005). But multiple studies (see, for example, Bloom, 1985; Dorner & Scholkopf, 1991; Ericsson, Krampe & Tesch-Romer, 1993; Holyoak, 1991; Hidi, 1990; and VanSledright & Alexander, 2002) have pointed to the importance of motivation and interest in the development of competence in any field, while Bereiter and Scardamalia (1993) see motivation as one of the key factors that pushes experts to work harder and more strategically than non-experts at both simple and challenging tasks within their domain of expertise.

This study examines four specific attitudes to writing: the extent to which scientists enjoy writing, their motivation to write, their resilience as writers, and their self-efficacy in relation to writing.

BELIEFS

> Further inquiries into scientists' . . . beliefs about writing are needed.
> — Yore et al. (2002)

Harding and Hare (2000) suggest that we have insufficient information concerning scientists' beliefs about science—and, we might add, their beliefs about the nature and purpose of writing (Geisler, 1994; Yore et al., 2002, 2003, 2004). In this study we focus on scientists' beliefs concerning the purpose of science and, more specifically, the role of the scientist in communicating scientific findings.

One of the major debates in the literature on scientists as writers concerns scientists' beliefs about the purpose of writing. Florence and Yore (2004), Bereiter and Scardamalia (1987) and Keys (1999) suggest that there is a distinction to be made between emerging and experienced writers, that emerging scientists see scientific writing as knowledge reporting, while more senior writers see writing as knowledge construction or transformation. In contrast, Yore et al. (2002), Yore, Hand, and Prain (2002), and Yore et al., (2006) found that senior science writers as well as emerging writers are likely to see writing as knowledge reporting. Yore et al. (2004) conclude: "the [beliefs of the] prototypical science writer . . . did not match the literature-based image. These [senior] scientists perceived writing as knowledge telling not knowledge building" (p. 346). Nevertheless, their findings are somewhat conflicting, and in their concluding observation in their 2004 study (p. 359) they note that the issue may be one of awareness:

> These scientists recognized the dynamic nature of writing as they applied the discursive movements to address the demands of the written discourse and the science technology content, but they did not explicitly use the language associated with the knowledge-building model of writing.

Other significant questions concerning beliefs relate to the persuasive quality of scientific writing and the significance of audience. Yore et al.'s findings (2002, 2004) suggest that scientists are highly cognizant of audience and the need to adapt their style and content to specific audiences. However, their findings relating to persuasion are again somewhat ambiguous:

> The respondents wrote to inform and share new ideas, unique innovations, and novel findings. Although none used the word "persuade" in their questionnaire responses, some scientists realized the rhetorical demands and the pragmatic need for writing grant proposals . . . [which] unlike research reports in which editorial referees can be persuaded by compelling

> evidence, must convince review panels by their use of linguistic devices in the description of the research proposed and its significance without the support of evidence (Yore et al., 2004, p. 357).

This observation is an interesting one since it distinguishes between persuasion which is achieved through "compelling evidence" and persuasion which is achieved through linguistic devices. This is a distinction which we will return to in Chapter 2.

One of the reasons why Yore et al. (2004) may have experienced difficulties in examining scientists' beliefs related to persuasion and the function of writing is that beliefs are often held tacitly, and therefore may not be easily identified through direct interviewing techniques (Benner, 2004; Dreyfus 2004; Tobin & Tippens, 1996). In other words, scientists may not consciously know what they believe about persuasion or the function of writing. This may be especially the case in a study such as this, where participants are answering questions about an aspect of their work that they do not consciously analyse themselves and where they have learnt a praxis intuitively (Florence & Yore, 2004; Jacoby & Gozales, 1991). While it might be argued that interviews and surveys are not as effective as other methods (e.g., observation studies) in eliciting attitudes and beliefs, I would argue that beliefs can emerge implicitly through narrative in response to indirect questioning and can be identified through close analysis of narrative texts. In this study, attitudes and beliefs were collected through direct questions (e.g., do you think scientific writing is persuasive?) and indirect questions (e.g., how would you describe the relationship between writing and science?), and identified through textual analysis.

Two final points about attitudes and beliefs: research suggests that they are developed early (Raymond, 2013) and they are resistant to change (Alger, 2009; Leavy et al., 2007; Martinez et al., 2001; Tobin & Tippens, 1996). This study therefore included questions about childhood attitudes and learning experiences related to writing and science: did scientists' early experiences, beliefs and attitudes to writing relate in some way to their relationship with writing within their professional life?

In this study, we examine four sets of beliefs—the importance of audience, the nature of persuasion, the perceived function/purpose of writing, and scientists' beliefs about their communication responsibilities as scientists.

The final chapter of this study, then, focuses on a model of the scientific writer which builds on current research into scientists as writers within the four quadrants of our model (learning to write, learning to write science, and the beliefs and attitudes of scientists as writers) to investigate scientists' perceptions of these issues. But should our focus be on the sample as a whole, or on the novice/expert divide (emerging vs. senior scientist), or should we be differentiating the sample in other ways?

DIFFERENTIATING THE SAMPLE

The current primary focus in the literature on the pre-post doctorate divide is perhaps understandable and in some respects desirable because, from the point of view of the writing community, the graduate years (and the undergraduate years preceding them) are the time when we can make a difference. As the doctorate is handed in, formal education is completed and there are fewer opportunities to engage across disciplinary divides. From the perspective of the scientific community, too, the handing in of the doctorate is the point at which a scientist's formal training as a writer, such as it is, ends (while post-doctorate positions do include opportunities for mentoring, the relationships between the senior and emerging scientists are generally less formal than that between the doctoral candidate and advisor). Focusing, therefore, on the final stages of the apprenticeship, where any formal influence is completed, makes pedagogical sense.

But while the current primary focus on scientists' perceptions of writing pre and post Ph.D. (novice and expert) scientific writers has pedagogical value, it also has limitations beyond those of disregarding scientists' learning experiences pertaining to writing prior to and following doctorate. A narrow focus on the differences between the two groups (novice/expert) may fail to recognise variation within each group (Carter, 1990; Dall'Alba & Sandberg, 2006): "Variation within groups could be considered an important research result, but in each study it was veiled by an emphasis on exploring differences at two distinct stages of development, namely novices and experts" (Dall'Alba & Sandberg, 2006, p. 390). The literature on doctoral supervision and co-authorship in the sciences does not distinguish the scientists who are engaged in supervision by experience, seniority, discipline or gender, or by attitudes and beliefs.

Yet even a casual engagement with the scientific community reveals that the writing tasks scientists engage with post-Ph.D., and the attitudes and beliefs they hold in relation to those tasks, are highly variable—though whether that is due to individual preference/attributes, disciplinary conventions, beliefs and attitudes to either writing/research or the responsibilities of the scientific community, professional life stage, or some other factor, cannot be established without empirical investigation. There are suggestions in the broader literature on scientific writing to support this observation of heterogeneity. Bazerman (1988, 1998), for example, suggests that competent academic writers in general are more inclined to write across disciplinary boundaries rather than focusing narrowly on the requirements of a single discourse community, and Yore et al. (2004) speculate that senior scientific writers may engage with a wider range of audiences, within the scientific community and in terms of the public discourse of science. Further, the preliminary analysis of the data for this study showed patterns of variation within the senior scientists which became an emergent theme of this research and warranted further analysis.

One of the questions explored in this study, then, is whether—and how—we might distinguish between categories of scientific writers. In particular, are post-Ph.D. scientific writers a single group, or can we distinguish between them in terms of tasks, attitudes and beliefs? And, if we can make such a distinction, what are the implications for both the scientific and writing communities?

The simplest approach to addressing the first of these two questions is demographic: can we differentiate the sample by discipline or gender? Differentiation by discipline was not undertaken in this study, since the sample was not structured in such a way as to facilitate comparative analysis.[5] Differentiation by gender, however, was possible—although any results must be tentative, given the smaller number of female participants. While the few studies that have examined scientists' or mathematicians' perceptions of writing (Yore et al., 2002, 2003, 2004, 2006; Burton & Morgan, 2000) have either specified gender when discussing participants' comments or included intentional sampling in relation to gender (for example, Burton and Morgan went to considerable lengths to find equal numbers of male/female participants), none of them have drawn any gender-based conclusions. Given that the role of women in science, and women's access to career positions in science are matters of some interest, this is perhaps a surprising omission.

A more complex approach would be to consider differences not simply by a novice/expert distinction but by "professional life stage" (Bent et al., 2007). Such an approach would be supported by the stage models of expertise (Dreyfus, 2005; Benner, 2004) which differentiate five stages on the path to expertise, each of which incorporates a differentiation in terms of skills, knowledge and affective factors such as motivation and self-confidence. Interestingly, such a model was raised by one of the participants in this study:

> So, at the beginning, when you start writing science, first of all, you're learning to write. Then you go through a phase where you are writing to discern what your ideas are as well as to understand your field and learn to write about science more effectively. And from there, you go on to writing reviews, where, in a way, you know what the ideas are at the outset. You know the field so well that you write and your

ideas are already formed. It's simply a matter of putting them out there. And then finally you get to a stage—or maybe not finally, maybe there is another stage after that—where you are providing a platform for something that has huge implications in lots of different ways. (Lemrol, Chapter 5)

While drawing only the broadest strokes, Lemrol's model is interesting, not least because it gives credence to the idea that learning to write science has a longer gestation period than has previously been assumed in the literature, but also because it suggests diversification of writing activities by professional life stage once a scientist has established themselves as a researcher.

Thus, one way to read the data in this study is to consider whether such stages can be identified in the "lifecycle" of the scientific writer and whether these stages can be further nuanced according to *affect*, i.e., attitudes, beliefs, and emotional investment, as well as behaviours and practices. This question of life stages was addressed in two ways in the study: the inclusion of a direct question within the interviews and an analysis of the complete data set into three categories (doctoral, emerging and senior scientist) according to tasks, beliefs and attitudes.

One final model was investigated to potentially differentiate the senior scientists. Based on a preliminary analysis of the data, which suggested that at the senior level we could distinguish two, if not three, groups according to task choice, I adapted Holyoak's (1991) model of routine and adaptive expertise to the question of scientific writers (Verschaffel et al., 2011). Holyoak (1991, p.310) defines routine experts as "able to solve familiar types of problems quickly and accurately, [but having] only modest capabilities in dealing with novel types of problems." Routine experts ("the artisans," Bransford, 2004) learn complex and sophisticated routines which they skillfully apply in familiar situations and, while they may continue to learn through their professional life, their focus is on increased efficiency at applying familiar routines. Adaptive experts ("the virtuosos," Bransford, 2004), by contrast "may be able to invent new procedures derived from their expert knowledge". For the adaptive expert, learning is about reaching beyond their skill set, embracing even uncomfortable challenges which will stretch their beliefs, and entertaining new possibilities.

Using this classification, routine writers might be seen as scientists who write consistently well in a specific genre to a specific audience (e.g., the physicist who writes consistently and exclusively for a narrow range of journals targeted at his or her disciplinary peers) within the established expectations and conventions of their discipline; someone who may, indeed, learn most, if not all, they need to know about writing during the postgraduate period apart from issues of efficiency. Many of the scientists in Yore et al.'s (2002, 2004) studies, who write for a narrow range of audiences, appear to fit this model. Adaptive writers, by contrast, would be continual learners, choosing to step outside of conventional disciplinary expectations and norms, adept at problem-solving strategies for discourse problems within and beyond their domain. Such a writer would need to continually adapt and renew their writing skills as they sought out new audiences and extended their skills to write in a range of genres to a variety of audience (for example, extending out into popular writing or social media), something for which their Ph.D. couldn't prepare them.

Holyoak's model therefore goes beyond stages differentiation, suggesting instead that there are different pathways *within* the senior scientific community. One of the questions for my investigation, then, was to consider whether we could distinguish these two groups within the senior scientist sample and, if we could, whether there we could identify distinct attitudes, beliefs and influences for each group. The preliminary analysis of the data set suggested a differentiation into three groups according to writing choices:

- Routine scientific writers. This group comprised those who wrote exclusively within the expectations of their discipline. This did not equate with those who wrote peer-reviewed papers only; an extension scientist, for example, might write in public-facing genres, but this would be within the expectations of their discipline. It also did not equate with people who avoided cross-disciplinary work; again, in some disciplines this is expected or unavoidable. The key differentiating feature of this group was that they stayed within the confines of their disciplinary *expectation*s. An example of the routine scientist is Mason in Chapter 3.

- Adaptive scientific writers. This group was defined as those who had intentionally developed opportunities to extend beyond disciplinary expectations through writing for public audiences, attracting cross-disciplinary colleagues as research partners or creating/finding opportunities for engagement in creative projects. The key defining feature of this group was that they had intentionally sought out research and/or writing opportunities that went beyond disciplinary expectations. Examples include the scientists in Chapter 2, or Gao and Elizabeth in Chapter 4.

- Transitioning writers. Although I had originally identified just the two groups discussed above, a third group emerged from the preliminary analysis: researchers who were either transitioning from routine to adaptive, or who wished to make the transition but were unable to do so for some reason. Paddy in Chapter 5, although an emerging rather than senior scientist, exemplifies this type of writer.

In the final chapter of this study we engage with the story of the data set as a whole, focusing on the model of the scientific writer as outlined in this chapter, and investigating the question of differentiation, and the implications of the findings as a whole for the scientific and writing communities. Before we engage with this larger story, however, we return to the idea of the individual story, the lived experience of the scientist as writer.

CHAPTER 2
REACHING OUT

> It's not done until you give it back.
> — Emerging Scientist, Ecosystems Ecology

> I put some of my best papers in journals that maybe would be considered second string [because] I would rather my work reach a dirty fingernails control person than a theoretical evolutionary biologist.
> — Senior Scientist, Ecology

The scientists in this study who "reached out" in their field, beyond their disciplinary field, fell into three camps: those who wrote for the public, those who chose to collaborate in writing across disciplinary fields or subfields, and those who wrote for a generally science-literate audience (ranging from undergraduate texts to journals that targeted a wide range of disciplines, such as *Science* or *Nature*). Almost all of those who wrote for the public believed that science had a strongly social, cultural, or economic imperative. Richard and Cameron's narratives in this chapter illustrate this point, in that both were passionate about the potential impact of science on the national economy and wellbeing, and both felt a social responsibility to communicate both the beauty and purpose of science to a community that supported them. Richard, in particular, argues something very similar to Randy Olson's (2009, p.8) comments about the public's attitude to science, possible implications of the anti-science movement, and the responsibility of scientists to do something to address a real threat:

> A backlash has developed against science, in disciplines ranging from evolution to global warming to mainstream medicine. An entire anti-science movement has emerged that truly does threaten our quality of life. Large groups of people are fighting against hard, cold, rational data-based science and clinical medicine. . . . Major groups are now arguing against certain childhood vaccinations . . . that have been responsible for eradicating terrible diseases. It is a genuine threat to society. In the midst of this conflict, communication is not just one element in the struggle to make science relevant. It is *the* central element.

Those who chose to write across disciplines were likely to be motivated by both a broader vision of the nature of science, and sometimes a desire to invigorate their careers in new ways. An exception to this is James who was one of only two participants who had started his career with two distinct (but connected) disciplinary interests.

For most of the writers in this study who reached across traditional disciplinary boundaries, writing for the public or across disciplines was an emotionally invested decision, although some, such as Marama, fell into a public profile almost by accident. Most saw it as a challenge, but a challenge that, for the most part, they relished. All of them felt it was far easier to stay within the narrow parameters of "what I know best." Some had difficult experiences of writing outside their comfort zone and described the need to toughen up in the face of public opposition or even hostility. There are consistent difficulties in balancing a sense of responsibility—of needing to communicate—while somehow managing the scholar's need to attend to detail. How to find a new, perhaps more speculative or persuasive voice? Despite the difficulties of writing for new audiences, most of those who did so saw the experience as enlivening their careers and being consistent with their own values and those of science.

There was, however, some disagreement concerning the academy's position on scientists' active engagement with the public. James suggests that academia is changing in relation to this, that "doing a Carl Sagan" is no longer frowned on but seen as a legitimate role for a research scientist. In Richard's view, the academy must do more to train young scientists to engage on the scientific-public interface in the interests of transforming public perceptions of science. Cameron, on the other hand, while himself seeing clear value in his outreach activities, suggests his peers sometimes implied that he was selling himself and his discipline short by being distracted by what were perceived to be tangential activities. Another participant expressed the dilemma as she tried to get to grips with why she doesn't do something that she thinks is important:

> I think that scientists, that we need to find ways and outlets to reach out to the public because—first of all, it's become expected of the granting agencies . . . you know, within this broader impacts category, how does your work make a contribution to non-scientists, to the lay public—well that's not all about what broader impacts are but it's important. Why am I pausing on this? . . . If I thought this was so important why haven't I done it myself? I don't know. Maybe because it's not valued by our institutions, you know. If you write a popular article, you don't get to count it as one of your peer-reviewed publications—well, it's not peer reviewed, right? So it doesn't count for much. (Senior Scientist, Biochemistry)

This is the question that scientists face: even if they know that communicating to the public is important, and even encouraged by funding agencies, can they afford the time to engage with such activities when they are struggling to establish and maintain research careers? As Cameron and Marama observe, writing anything but peer-reviewed papers may be a distraction for emerging or developing scientists who have not yet established their position in their field, and there are political risks associated with taking a public position. The four scientists in this chapter, the most prolific public-facing writers in my sample, had well established research careers and, in a sense, as Marama says, nothing to lose. Yet, if senior scientists are most likely to have the freedom to engage with public discourse, as Richard observes, what impact does this have on public perceptions of science?

The figures for the overall sample in this study showed almost no distinction between the numbers of senior and emerging scientists engaging in public-facing writing (35% and 34.5% respectively). However, there is a clear distinction in terms of intent: the emerging scientists who were engaged with public-faced writing were most likely to be found in an applied discipline where writing for a specific public audience (e.g., health professionals) was expected; the senior scientists who engaged with public-facing writing were more likely to be targeting a wide general audience on their own initiative. Although several female participants suggested that there were more pressures on women to be engaged in public discourse, this was not borne out by this study.

One of the central challenges for those who engaged with public writing concerned the issue of persuasion. Many of the participants in this study were hesitant in addressing the question of whether scientific writing was persuasive—a finding that is consistent with the findings of Yore et al. (2004) who noted that while participants in their study were most likely not to see their writing as persuasive, they nevertheless used persuasive techniques in their writing. Most of the hesitance around persuasion related to participants' conception of persuasion as equating with emotive language or speculation:

> In certain types of articles there [is a role for persuasion]. But they tend to be more the popular press. And some conferences—when the proceedings are published—they'll give a particular avenue for more opinionated pieces of writing to be published. But for pure science writing, no. There's no place for speculation; if one is speculating, it's poor science writing. And if one is trying to persuade, as opposed to lay out the facts as they were in an unbiased impartial way, one is not writing science. One's writing something else. But it's no longer scientific writing.

Chapter 2

> I suppose it comes back to what we mean by "persuasive." Scientific writing should be about interpreting the facts as they present, and saying no more than that. Where persuasion tends to infer that . . . persuasion is often biased. It's often saying more than what you can actually glean from the facts. It's taking a position. There's no place in science for taking a position. The facts take the position. You must simply present the facts. One doesn't have a role in speculating, in trying to read something more into it than there actually is. It's very, very important. That's at the nub of science writing. (Senior Scientist, Nutrition/Physiology)

This was a recurring theme: scientific writing is not persuasive; it's just laying down the facts. And while we might question this, and argue that scientific writing is also about interpreting the facts in the context of prior research, making meaning—"telling a story" as so many participants described this process—nevertheless the point is made that in scientific writing the researcher must stay close to their data, laying down the evidence with, as James puts it in his narrative, "my 10,000 footnotes over which I laboured lovingly."

For those who step across the peer/public divide, though, the issue of persuasion takes on a new complexity. Separated from the familiar routines of laying down evidence, how do scientists write persuasively and compellingly, while remaining credible, for an audience whose knowledge is limited? All of the narrators in this chapter address this issue: James finds a way to write his popular articles so they "have the feel of a thought piece where one tries to build in 'this is on my mind, I'm curious about it, this is a trend I see, let me try it out' as opposed to 'I will now prove this beyond the shadow of a doubt.'" Marama has been bitten by public comments and has learnt to simplify. Richard talks of sifting out the detail to tell a simple story. For Cameron, crafting a message for the public is about having really clear aims and a targeted audience.

I have chosen, for this chapter, the four participants who were most broadly engaged with a range of audiences. All four are highly successful senior researchers who have engaged in cross-disciplinary collaborations and the public discourse of science. They are domiciled in a range of locations (the US, UK, and Australasia). They are all, surprisingly enough, physicists. And they were all, it seemed to me, people with big visions, and hugely flexible writers, with a strong sense of need to engage the public with the possibilities—and the wonder—of science.

JAMES

James is a busy man. He publishes widely in two distinct disciplines, in a wide range of genres, and as well as this holds a number of administrative roles including chair of his department. "How do you have time to sleep?" I wondered towards the end of the interview. Yet he appeared, when I spoke with him, to have all the time in the world. His office is a model of organisation and calm, the bookshelves that line every wall filled to capacity in an orderly fashion (though I'm sure I saw *A Wrinkle in Time* sitting next to a book on military history). Because he writes in such a range of genres, and in different disciplines, he has a valuable perspective on writing and disciplinary specificity—and on the value on writing for a public audience. And it's interesting to note the number of times he describes "fun" as his key motivator.

IT FEELS LIKE A FREENESS TO TELL BIGGER STORIES IN MUCH SHORTER SPACES

I was trained in both theoretical physics and the history of science and I teach here in both departments. As a young person I was very excited about science and was reading a lot of popular science books. Some were very old, from my father's bookshelf—books that had been popular in the 1950s—the great, physicist popularisers of some time ago. When I got to university as an undergraduate student, I was convinced I wanted to study physics but I was also fascinated by the human story of science from these books; many of these books are very kind of heroic, you know, almost hagiographic. Luckily my first week on campus one of my physics mentors said "there's this thing called the history of science and it's actually more interesting than only great stories of great men—there's actually a broader human, cultural story to be told." And so I decided to pursue both at graduate level. That's the short version of how I came to try to have a foot in both worlds.

It was much more recent, though, only really since I began teaching here at my current institution that I began trying my hand at other genres. I had been trying to publish fairly straightforward academic history in recognised scholarly journals, and likewise in physics journals. And then after a few years of that I had opportunities to try to write for other audiences with different scales—short popular essays and nowadays little bits like blogging. And I really enjoy it. I enjoy the challenge of the shifting genre.

It's fun—it's new and it's hard. I find it difficult, and that's a fun challenge. I've learned how to write the 40 page double spaced academic paper for my history colleagues and I can get better and I still enjoy doing it, but that's a recognisable exercise. And likewise with physics research articles. But it was something quite new to me to try to write in 1,800 words or 3,000 words for an audience that will not share the same background as my usual audience.

The first feature article I wrote was for *American Scientist* and I was advised that might be a good one to try a toe in. Its readership is mostly other Ph.D.s in the sciences but across the sciences, so it's a certain kind of writing for non-specialists, writing for highly educated science-savvy non-specialists. And I was writing a 3,000-word version of the long book I'd just finished in the history of science, a 400-page monograph. That's a non-trivial task. There's a lot of technical material on quantum field theory that I was grappling with while wearing my historian hat, thinking about how art historians think, about artistic styles and representation. I had just finished the manuscript and the book was probably in press at that point, and I was immensely frightened about all the things that I imagine are familiar. I can't have footnotes? How do I show the reader that I did all this hard work and I'm speaking from some sort of authority on the topic? And that was a very brisk learning experience.

So it was not a scholarly style, not the scholarly apparatus of, "here's my 10,000 footnotes over which I laboured lovingly." This was also the first time I'd been heavily edited by a very skilled editor. What passes for editing in scholarly publishing tends to be rather hands off and minimal. In physics journals there's virtually none, and so to have a very smart editor push back and rearrange and pull this out—she was very good—was a good lesson for me. I must say I've been lucky with mostly very talented editors—at the *London Review of Books* there's one person I tend to work with whose edits will often surprise me and my writing is always the better for it. So I've developed an appreciation for the art of editing, especially these very short pieces for very broad heterogeneous audiences. That's its own separate skill it seems to me.

My most recent book was with a trade press aiming for a broader audience. Both of my books were single-author monographs of comparable length, but certainly quite different on the page and I was much more attentive, or trying to be much more attentive, to character, to the plot. I was trying to get a lightness of touch there. The title's meant to be very ironic and not to be taken at face value. The picture on the cover should prep a reader to think "this is done with some sense of irony."

It was an experiment for me, again a learning process, to see could I first of all try to explain some of the ideas in quantum theory that are very difficult, and very technical, and they sound like magic even when other physicists try to describe it. And yet they have a kind of beauty and power and have come to be enormously influential, they're really big important physics ideas that I really like, they're neat.

I'm trying to get across some very hard ideas within the history of science about knowledge being produced in very specific cultural institutional settings, that it's not free floating among great minds, and yet for all that, it's not just made up, it's not just willy-nilly. I'm trying to convey an idea of a kind of cultural embedding of scientific practice without sounding too ponderous about it.

And then frankly I also wanted to tell a good story. These people I stumbled on who I talk about in the book have led remarkably interesting colourful lives. Many of them are really out of central casting—they are just larger-than-life personalities. So I was trying to get some basic storytelling about the characters and the plot, with which I really struggled much more than with my earlier book. "I know how the story fits together, I know my argument, I know the pieces. So what do I do about the reveal?" I had to think about when to lay this out, in the sense of foreshadowing, so I could come back to it. I had not paid nearly so much attention to that in my earlier academic writing.

But I'd say the centre of gravity of my work has remained the bread and butter academic publishing: the physics articles should look like physics to the physicists; the history articles should look like history to the historians.

I decided to write for a more popular audience partly because it did seem like it would be a new fun challenge, because I knew it wasn't just the same and there were many science writers who I so much admired. Also I came to a point where I realised this stuff is just amazing to me—both the physics and the history. I mean, that's what keeps me up at night. And I thought that needn't be merely the province of a small number of other specialists. I had grown up reading popular science of a certain kind and I really enjoyed that; I know how much that inspired me, just made me interested in things I hadn't thought about before. There is this enormous smart, educated reading public. And that's when I thought it would be a worthwhile challenge for me to try to write, to engage with some of them.

I think there are some very clear differences in writing for history and writing for physics, at least in the writing process. I'm sure we could dig deeper and find commonalities. When I was finishing graduate school, I wrote two separate dissertations for two committees, precisely because the genres were so distinct; I didn't want to have the physicists throw up their hands and say "what is all this verbiage, what is all this crazy history stuff with all these footnotes?" And vice versa. Many historians on my committee would have no interest in, or the kind of training needed, to read these very specifically crafted, highly equation-filled physics ideas. And so with the history dissertation, it obviously requires years and years of running around archival sources and interviewing people and digging up all these old dusty things from many places. That was the research phase and it was at least as long if not longer in time to figure out how to write it.

But I felt almost no separation between doing the research for the physics and writing it up. I used to joke there was no trauma of finding my authorial voice around my physics dissertation—I didn't have to worry about meeting my audience halfway. With the physics, it felt very much like the writing was of a piece with the research. I mean, writing physics already felt like a default.

In the history of science, on the other side, we have to worry about historians of science who don't know anything about quantum mechanics. But when I was writing up my history dissertation, much of it went into my first book, I really struggled with a shadow of what I struggled with in my later book: "what does a reader need to know right here?" and "what do very smart historians who aren't specialists in this very specific thing need to know to get them ready for the next historical point?" That requires a level of attention to plot, almost like a detective story; the reader must know this, here, to lay the seeds for what comes later.

I didn't feel that when working on the physics dissertation. I think of that more when I'm writing my physics articles now because there are, of course, better and less well composed physics articles. One still wants to bring the reader along. So it's not oil and water, but I felt it as ends of a spectrum. It's a matter of degree and not of kind I would think. Writing my physics dissertation felt like almost an automated process. First of all I wasn't expecting many people to try to read the physics dissertation, and the physics dissertation as a genre had largely become a bunch of articles that had already been written, so it's not that one needs to plot out a many-hundred-page through-line or a narrative arc. It's not sustaining an argument, it's not sustaining even really particular themes; each chapter I think is treated much more autonomously in a dissertation. And within each chapter, I felt like there's a much clearer expected flow or outline; I could almost plug things in. Of course there's an art to that and some people do that better than others, but it didn't feel as wide open. And so it really felt like, as soon as I figured out my equations and my computer simulations for the dissertation, I more or less had written it up; I had done so much of the writing itself, it was just stringing together my derivations and my equations which I had in my notes and trying to tell the story to bring a reader along. But the expectation would be that the reader was already so hyper-specialised and following the larger story already, that I didn't feel the need to bring them in and orient them and help them on their way.

I think empirically that reading practices have changed enormously; it would be fun to do a more careful study. I think now physicists have moved not just to online reading but to one homogenised central server, then much more reading is done on the fly, online. And so what one imagines is people really just read in a continuous scroll through equations and then sometimes they will indeed print it out and read a bit more carefully. So you can see that one aims to see in two clicks whether this other group is studying the same model as mine or not. I see "oh, here's how they define their model" and I can see bang, bang, "OK they're doing this, they've left that out, they're putting these effects—OK, now what do I do?" There is a level of a much quicker visual scan. The words will matter but they'll matter in a different order and people will routinely just skip the introduction.

Sometimes the introduction is written for the co-author—I say that partly as a joke—to make sure that two or three people are agreed on what the goal is. I work very hard on my introductions in physics papers. Often these things are read in many ways by many different types of people. So they're written for the referees, they're written for the very specific fellow specialists; partly it is a way to mark turf and say "this important work has been done here, but I'm not just doing what they've done." There's a kind of differentiation that goes on there. And that might be aimed at very specific colleagues or even competitors in the subfield. And again, you know, some journals will put a premium on purported novelty and originality and so again one broadcasts that in the introduction or the abstract even for the kind of workaday journals like the *Physics Review*. Introductions are important, but they're not important in the same way for all readers.

History papers, on the other hand, I tend to read quite linearly. I certainly start with the introduction and pay much more attention to my own, as well as to those of my students or colleagues or a random piece in the scholarly journals. I start at the beginning. Sometimes what I'll do is read the introduction to get the orientation: what are the main themes, what are the stakes? Then sometimes I just skim the footnotes. What are the types of sources? What are the sorts of names and dates and places that are likely to come up? And then I go back and read the body of it.

I like writing history and physics. Honestly, I enjoy the contrast because they do feel so different in construction and in reading them. That's part of what I enjoy—it's a constant reminder that there is a complicated world out there that we carve up and try to understand with many different tools, no one set of which has conquered the world. And likewise there are many ways to try to capture the complexities of the world in something like narratives. And so I find that fun; the contrast is enjoyable.

It's a similar enjoyment in the newness of a style, like trying to write the short popular essays. I often find it very hard to say something big in an academic article, at least in an academic history article. Ironically, I feel like we often force ourselves to write quite narrowly to our case at hand to sustain what often are, honestly, quite narrow arguments about small bits of space and time—people and places and times—so that our fellow specialists through peer review don't throw a fit or just because we don't want to overreach. We studied this bunch of documents with great care, but that doesn't mean that the story generalises. Whereas I think it feels like a freeness to generalise, to tell bigger stories in much shorter spaces without the footnotes, because the expectations are quite different, both of writers and of readers. I feel like I can sometimes say bigger thoughts in a *London Review of Books* piece, in 1,800 words, than in a very carefully, hyper-attentively constructed piece for a journal like *Isis* or the *Journal of American History*.

I certainly do write with some care not to dramatically overreach. On the other hand, there are trends, there are big ideas that are worth at least sitting with, and I think they can be tried out in a way that needn't come across as definitive and beyond question—saying more like "what about this?" They can have the feel of a thought piece where one tries to build in "this is on my mind, I'm curious about it, this is a trend I see, let me try it out" as opposed to "I will now prove this beyond the shadow of a doubt." So one uses caveat words carefully, oftentimes you don't have to make every statement an absolute. I have become progressively less fearful of that, but I felt that very much with my first few pieces.

My schedule has become quite chopped up, which has been a real frustration. Particularly because I used to be the type of person who would need, say, two or three days in a row to get any quality writing done. I felt like I needed that, and that simply doesn't exist anymore. So I'm lucky if I can devote a single work day to a writing project. And once I get going then I find I can work in shorter stints; often the start of a new piece of writing requires a day, or the better part of a work day. And then I'm getting a little bit better at filling in a choppier work schedule downstream.

I think through difficult issues by trying to talk them through, and often the talking through is as much to see if I had figured it out, whether I have cracked it or not. But it's also to try to think about the ordering. So, you know, "what does the claim really depend on, what does my interlocutor need to be on the page with me about?" Even if that's just in a conversation on Skype, it often becomes as much about "does this conclusion hinge on *this* being in mind before *that*?" It already starts getting into the process of persuasion I guess.

I tend to write with multiple word documents open at once, so I'll often start with a very scattershot outline that's sort of an outline-brainstorming that I can cut and paste around. Then I'll have a scraps piece; I'll start trying to write a particular portion of that outline in a new document and then come to realise "oh this whole part doesn't belong here, but I don't want to quite delete it" so I'll put it in a kind of receptacle file. So you would see me hunched over my desk alone much of the time, but with a few different files open at once.

I just wrote a short piece for *Nature*, 1,800 words, a historical comment piece, not a research article. And it actually felt rather freeing. But I tend to over-read; I found myself reading whole bunches of stuff that I knew would play absolutely no role whatsoever when I finally sat down to write that short piece. I felt the need, mainly because it's such a short piece and it goes back to what we were saying before about accuracy, authority, comprehensiveness, the kind of anxieties that came up about that. And so I found myself pulling things off the shelf that either I had always meant to read but hadn't yet or had read years ago. And none of that mattered; it was psychological. Maybe it was playing some orienting role for my thinking even though one would never know it from the very short piece that emerged. I guess it's a form of immersion. And I feel the same even when I'm thinking back to my recent physics research articles with some of my students, that again there's so much more that one often knows than goes on the page.

I do a lot of revision, certainly. And I tend to find—this is more true I think about historical writing—themes and anecdotes or episodes come into my first draft, and then I will work to see if there is an argument there, or what might my argument be. I'll get a story stuck in my head which is different from having arguments or theses, you know, and I start from that. It's not exactly gathering lots of historical data and then reaching my conclusions, certainly nothing like that. It's haphazard. I get intrigued by people or stories or episodes. I love to work that in somewhere: what role might that play, what would it be building up?

For example, the story at the beginning of my most recent book originally appeared in Chapter 5 in my first draft because this is the chapter where I talk about this theme, and that's where I want to introduce this main person, and so in my head that was contained in the middle of the book. And it was a lot later, in a downstream revision, when I realised I needed some hook for the readers. That little story also sets up the types of people we will encounter, and so again I began to think about foreshadowing. But I didn't think like that at first, and now maybe with that experience I might do that more intentionally. Something I have been doing in even my academic history writing for some time is trying to open with a kind of anecdote and then by paragraph 2 or 3 saying something like "the thesis of this article is"—you know, the more standard declarative approach—but trying to have even a one paragraph colourful little entrée. Doing that at the scale of a 30-or 40-page article is one thing and I hadn't really made the leap at the book length until this recent book.

Would I call myself a writer? Only recently. It's something I aspire to, something I would enjoy getting better at. And I say that because there are many writers whom I just admire so much. So that's fun; it's nice to have, you know, goals or models to aspire to and I find that very helpful. I'm a devotee of the *New Yorker*, for example, and I have my favourite authors within that constellation. And some of them will get me every single time, whether it's to laugh or cry or just an "aha!" or to say "now that's a writer!" Or I'll pause over a paragraph or sentence. And so I don't consider anything I've done like that by a long shot, but I have enjoyed trying to figure out how they did that, trying to appreciate the craft of that more consciously. That's also why I continue to be very delighted when places like the *London Review of Books* or *Nova* have come back to me, now several times, to try to explain things for this or that group, because I perhaps have some more experience of that than other people. That's a skill that I enjoy trying to cultivate.

I do enjoy writing. Yes. And, you know, I enjoy going back to things I've written long ago. I can remember and describe at various levels of detail what my first book was really about, or even what the more recent book was about. That one has been out for two and a half years by now and there's a special pleasure in serendipitously going back and just opening up and reading at the paragraph level, not the big ideas, not the main thesis and not what archives I had to visit. But I'd say "oh that was really fun"—the joy of crafting again at the small scale. I remember now beating my head against getting that turn of phrase right, but I'd forgotten in the interim.

I know people who wrote a textbook as assistant professors or even as postdocs—though by no means is that the norm. So that's different from the end of career, legacy model; and these were often astoundingly successful textbooks and sometimes they were trying to start a new field rather than cap an old one. There were other reasons why they chose that path than the usual one. Likewise, I know many, many scientists who are actively blogging and doing some kind of writing for more than just narrowly construed technical peers. And that really wasn't true to the same extent 20 years ago.

These changes are not happening in a void; at least here the deans tend to be, in my experience, very pleased when even their very young faculty do something that now they'll call outreach—now that's a positive value, instead of saying "oh you sold out, you're doing a Carl Sagan," which was taken to be a very bad thing 30 years ago, unfairly. That is not unrelated to the shifts in national priorities and science funding, and debates over the place of science in politics and culture and so on, school boards denying evolution. I think there's less of the immediate instinctive reaction to say "stay in your ivory tower, do real science." As an example, the dean of humanities and social science a couple of years ago sponsored an event aimed at faculty and even young scholars, grad students, on how to write for a popular audience, and she brought in literary agents and editors from the significant news outlets as well as trade presses.

I began publishing physics research articles as an undergraduate—the first came out when I was still a student and then a few others based on my senior thesis, so I was trying to get habituated in the rhythms of writing for that kind of audience. That means learning how to use things like passive voice and learning to have a certain kind of standardised structure.

I learned to write by immersion; I had to be immersed in a very specific collection of articles on the same topic just to get up speed for the questions I wanted to ask and the calculations I wanted to tackle. At the same time that meant I was reading samples from a very specific kind of literature, reading them over and over and over again. These days I think we've collectively gotten better at building that stage into undergraduate coursework. Not for every course, but for several of our physics major courses here, the students will write a version of a research article. They'll learn to use LaTeX, but they will also learn the basic features that would go into a physics article and they'll often have to write a final paper that is in the recognisable format that would look like something I read in the physics journals.

I don't recall doing that as an undergraduate, not as directly, not as part of a built-in part of the curriculum; I'm glad we do more of that today. I had a bunch of independent studies as a student; I was working on a senior thesis and I had opportunities to practice more one-on-one with faculty mentors. But I remember, what became my first physics article was the second version I submitted to the journal, and there was an enormous transformation between those two drafts; the topic was the same (in my head at least), but if you could do page-by-page comparisons they would look very different.

I think the biggest shift was the notion of what's the novelty, what's the contribution as opposed to summarising what I had learned from other papers. I think a lot of the equations might be the same, I'd have to go back, I don't remember. But the first draft was, I would imagine, more like a book report—here's what I've learned, and there was some new stuff in there—but it was not pulled out or not placed the way one might try to do that. And then several months later, and with lots of help from advisors, I was able to firm up "what am I actually trying to say; what do I think is new here? Why is this relevant and interesting?" And then to place it in relation to what had come before, without making it a kind of continuous story of what has been known. I bet that's the biggest difference between those two very early, very brief papers.

In terms of who influenced me as a writer, going back to undergraduate studies there's one advisor on the history side who was a delightfully picky grammarian—she insists to this day on very clear prose. She's a very gifted writer as well as scholar. Though her own style was not exactly *Strunk and White* pared down to a stereotypical minimum, she showed me one can use interesting and colourful concepts and theories, but the prose needn't be flowery. And to get that straight in my own head was an important distinction. She was very, very helpful with that—draft after draft after draft—she was very generous with her time. So partly it was "what are you really trying to say?" and answering the "so what" question. My colleagues and I often say with our history students, "well yes, you just dug in these archives and interviewed these 50 people or whatever it is; so what? What is the question you're trying to answer? What's the argument going to be?" And then you can worry about how do these blocks fit in to sustain that argument.

So it wasn't merely at the level of copy editing with split infinitives or commas; it really was an architectural metaphor. I think there was similar stuff going on with the physicists, that was probably articulated less explicitly.

I do think I have a role as a teacher of writing. And I take it at least as strongly or more so on the physics side than on the history side, because a lot of teaching writing is built into the process of becoming a humanities student. We have an infrastructure to help students work on rough drafts, to revise constantly, to worry about authorial voice, something like a plot. So that's on my mind when I look at my own students as well, especially my graduate students who are looking at published things. But even with our undergraduate essays, they have to worry about a thesis statement, an argument, which is the basic structure of essay writing, and that's built into what they're getting when they take a history course. I'm now publishing with a number of my undergraduate physics students and Ph.D. students as co-authors. And I try to be very attentive to helping them to guide the reader along and not letting them get away with an attitude of "you got the equations done so now you're done," the way that, frankly, I had largely felt as a younger physics student.

And there's another thing I'm curious about: there are enthusiasts or zealots or ideologues about online-only hypertexting, saying this is the natural medium. I'm not sold on any of that; I think reading books from one page to the next actually has worked pretty well for a couple of hundred years, so I don't mind linear constructions. But I'm curious about the affordances. I'd love to write a shorter ebook on a popular physics topic with maybe some historical framework, and then to have the words, metaphors, pictures version of some important physics concept embedded in it. And there could be these layers. Let's say someone has taken high school calculus, well there's a version for them, and then maybe a third layer for someone who's a bit more advanced than that, with the technical stuff linked in. So, not to write a textbook and not to merely reproduce the physics specialised literature, but there is a way of trying to say "this idea is even cooler—have you noticed, have you thought about this very subtle thing" that's expressible in a certain medium that might not be essential to the main point that I would hope many readers might otherwise be able to get at. So I do think this kind of non-linearity would be really interesting to play with.

I really have been interested, partly as a participant, but often as an observer of this fairly recent shift to really encourage or at least no longer overtly disdain so-called "real scientists," with a day job in science, making increasingly creative forays into creative non-fiction, from blogging to sometimes very successful books. And I find that a wonderful transition; there's a notion that many ideas in science are cool and they're exciting and they're hard and some of them are quite consequential, and why on earth wouldn't we be trying to at least talk about the excitement of it? Not as a gateway to create legions of other specialists, but rather to say "this is an essential part of human culture and it's a grand intellectual adventure that has a rich history."

I wrote a short little physics article just a few weeks ago with two younger, up and coming physicists, and it was a great joy, really fun ideas. And we worked very hard, I worked very hard because I was learning a lot from them about specific things I didn't know about and it was cool. It's a fun and juicy topic and I was just excited about it and then we posted to this physics pre-print server. And about one or two days later we found another physicist blogger who wrote a whole blog piece about it and was so excited he mocked up his own Photoshop images about it just because he thought it was fun and then wanted to blog about it.

So there are these physicist communicators who are reading the pre-print server religiously, routinely, and are actively, in real time, communicating that "this is cool, exciting, here's why I think this is the most fun, cool, exciting idea." So the medium has allowed a kind of fluidity which I like.

I'm just not a blogger by nature, but it's fun to be able to point in the hyperlinks to my colleagues' hard physics papers that are freely available on the web, to say, "look to learn more about this, at least this guy has cool pictures in his paper, you don't have to read a very dense technical paper but I can try to give you a paragraph of why I think it's so neat and why it's interesting and why in the light of the latest results from some experiment we want to rethink this older paper." And I can at least speak to the real heart of the matter, the real article. There's a real fluidity that many eyeballs might be hitting on any of those and following from one to the other. So it's not that there's a brick wall, and the real physics or the real science is over here. And I think that's great.

MARAMA

Marama is that rare thing: an eminent female physicist with a high media profile. She speaks here about falling into a public role by accident, by trying things out and seizing opportunities. And she talks about the risks of having a media profile: the risks to younger scientists, the risks of alienating colleagues and institutions, and of managing hostility from a very different audience than that to which she's accustomed. But while she sees her writing sometimes as a burden, she also experiences writing in new forms as a kind of liberation, of finding a new voice inside her that was longing for expression.

I Finally Found this Voice that I Didn't Know I Had

I never decided to take on a public role. It seems to me that I've risen up through the ranks and become a professor of physics, and over the last decade I find that I've got this other increasing public visibility, which I did not set out to achieve. So it's a question of, since I have this visibility, trying to think how to use it wisely and effectively for the things that I care about.

I think there might be three reasons why I got into public writing. One, I'm a female physicist. There aren't many of us. The choice is more limited than if they want a male scientist. Two, I think a lot of it does comes down to the fact that I started blogging. And I did not start blogging to have a scientific, or a public profile, nor did I start blogging with any clear aim, if you like. I was just encouraged to do it by someone who was themselves blogging. I realised that I was reading more and more blogs and wanting to comment, and he said well go on, give it a go. I guess over the first few months I found that I had a voice. That there were things I wanted to say, and that there were people who liked what I wrote. And that encouraged me to do more. Then, the third thing was that over the next couple of years I got asked to write a few other things. It just built up from there. Of course, a lot of it is because I write not about science per se, but about the social and cultural context of science. I'm one of the few senior women who is speaking out about it. There are many early career women who are talking about these issues, but I'm coming at it from a different direction.

The thing that perhaps surprised me is that, when I started writing my blog, I felt liberated by it, because instead of having to write in the passive voice in a very formulaic way, where clarity was important but not necessarily creativity, I have found it great fun to have this other voice.

Someone said of me quite early on that it looked like I had all this stuff inside me that was just waiting to come out. And I think to some extent that's true. I felt that there had been a part of me that had been suppressed, which I wasn't aware of. While you're getting on with everything else in life you're not necessarily aware of what you're not doing. But it was as if I finally found this voice that I didn't know I had and it made me feel more complete.

I am very conscious of the fact that I write in different styles for different places. I do write sometimes for the mainstream newspapers and that has to be very different. The first time I wrote a relatively big piece for a national newspaper, I got completely hammered by the comments. It taught me that I tried to put too many ideas in 500 words or whatever it was. It, therefore, confused people because I was mixing in different ideas. So that was very salutary. I have my own blog where I never really write about science, per se, but when I write for a national newspaper's science blog I will sometimes write about pure science, take a paper and analyse it. So that's different again because you're trying to take advanced physics and turn it into something that the general public can read.

When people attack you, and when people you know disagree violently with what you say, that is very hard. It doesn't mean I don't feel I shouldn't have written it. But writing things badly from experience and finding that people can pick holes in your argument is upsetting, and occasionally one makes mistakes—I have done some really daft things in my time. But there are also times where you just know that people will never agree with you. And writing about social issues in science, there have been people who just consistently attack you for the very ideas. You can despair.

Scientists don't talk about writing because it's not seen as a core competence. I'm of a generation which did get proper grammar teaching at school. I have always worried about how one writes. So my students, say drafting a paper or a Ph.D. thesis or whatever, they are surprised that I will correct their grammar and give them ideas about stylistic stuff. I've always done that. It seems to me absolutely fundamental that one communicates. And if you write bad grammar or in a rambling way or your sentences don't have a verb or whatever it is, it matters. But students think that the only thing that matters is the result and they don't understand that communication in the broadest sense is important. They feel "I did the experiment right and these are my conclusions." So I think it's part of our education. We do not value writing when we train scientists. So if you're a historian, you are instructed on how to write essays, and if you are a scientist, you are instructed on how to solve problems. I think many academic supervisors probably themselves don't get it. They play it down. I think writing matters far more than people appreciate, and I think we fail our students, our undergraduates, in not making that clear. Because they will have to write even if it's purely an internal report for their company when they do a piece of research. They have to be able to communicate.

I think creativity is an aspect of science and, therefore, of course it's going to come into an aspect of the writing. I think people often ignore that. I get really frustrated by people who say—it's the old Blake idea that science is just measurement and it destroys wonder. I think—there's no creativity; Blake was of course not a scientist. But I think there is this belief that science is non-creative. Science is very creative in the way you tackle a problem and, therefore, that should be part of the writing process. But, and this comes back to why I feel liberated, I think it is true that if you're writing a scientific paper, that's not the primary focus. You're not being creative. I think the creativity comes elsewhere. I think one can be much more creative in other kinds of writing.

There is some narrative craft involved in writing science: that's absolutely right. Depending on the kind of results that are being written up, it may or may not be obvious how to do it, but you certainly have to think about it. I think one of the things that I know I initially had real problems with is getting beyond the chronology. It's very easy to try and write a paper in the order in which you actually did the experiments. Because that's following the thought processes you had, and sometimes you need to step back and realise that isn't the narrative that makes sense at the end of the day.

With my blogs, I will start with an idea, I will think about the points I want to make. But it's not really until I write it down that I really work out what the narrative is. It's very easy to write paragraphs and then move them around to get the flow right. In those cases, I will know what the big picture is, the subtopics, the anecdotes before I start writing. And then when I write—to be honest, when I write I typically just then do a brain dump and I rarely have to do a lot of reorganisation.

I think you have to learn the style almost by osmosis unless you're going to have a formal teaching course, which at my stage of my career I'm not. Take Twitter—again there are stylistic things and very little of it is written down and you learn by watching. It's like being immersed in a foreign language. One of the things that I find surprising is how students don't manage to do this. Like they don't realise that they need a bibliography or they don't know how to write a figure caption or something. At the end of their Ph.D. they still seem completely naïve, although they've been reading papers all the time. But then I've never instructed them, "analyse the papers you're reading for the correct structures." Clearly it's not obvious to other people that there's this in-group way of doing things. I get draft thesis chapters where someone has completely failed to understand the point of a caption. And you say to them, "you just can't do it like this. Either you have copied great chunks of text into caption or you have just called it figure three." I mean, come on! Eventually they get there.

Time is a difficulty. Sometimes when you're trying to write—not a scientific paper—a blog or something for a newspaper, you want to be timely. That is a real problem in an academic's life. I think there are things that I would have liked to comment on, but there's no way I could and keep it topical. But I enjoy writing. Absolutely. I worry that I spend far too much time on it now; that it's taking over my life. I have all the other things I have to do.

I think the lifecycle model is certainly how I would see myself. Other people I know who blog do it more explicitly as part of their career goals. I know plenty of people who would fit that model, but if you look at an earlier stage in their career you can't really tell if it's going to be valid. I think it is difficult if you are 25 to start pontificating about policy because you haven't had a chance to absorb all the nuances. And a lot of people who do an awful lot of writing early in their career often feel that disadvantages them because they are writing and they are not beavering away at the bench to get the results.

There are some people who are very focused and have spent their entire career working on whatever it is. And then there are other people like myself whose research has evolved and not gone in a straight line. Sometimes I think it's about confidence. Not confidence in the usual way it's defined, but a willingness to step outside the area where you have a reputation requires that you have a certain degree of confidence. I've thought a lot about this because as a young person I would never have said I was confident. I would still say in some senses I wasn't confident, but I am prepared to try new things.

So it's about risk taking. It comes back to getting negative comments about what one writes. I have written really risky things and have been explicitly told "that was brave." You have to do that. And sometimes, I have been compelled to speak up despite the fact that it was absolutely not what people would expect me to do, given my background. I feel I have to do it for personal reasons, if you like. My sense of integrity. But it is taking risks. And sometimes you don't know what you're doing before you get into it. I think when it comes to writing or when it comes to research not everyone is prepared to do that. So other people might feel "this is my niche. I am very comfortable here. And this is where I'm going to stay."

It is important not to assume that people do things consciously. I think how you act is often circumstance. And also you to have to convince people there is something in it for them. If you're an early career researcher worrying about early career progression it can be a problem. A study that came out just recently about people who get involved in public engagement showed that they felt it damaged their careers and they didn't get encouragement to do it. Again, that's why it's easier for me as a senior scientist. If you are an early career researcher, you're trying to get the next position. You're trying to get tenure. So I think that however much you might want to take risks, you may not feel that you can, particularly if no one encourages you. If someone says, and I've heard this of people, that they might be more likely to get a job because they were perceived as having these other interests, then fine.

But I think you can believe something very passionately yet not feel it's your responsibility to talk about it. Not talk about it publicly. You do have to worry about the ramifications of taking a position. There are people who are writing about areas of research which become very controversial. Do you want to keep your head down and write a scientific paper or do you want to write for a broadsheet? I think integrity can come in many shapes and forms and just because you're prepared to speak up publicly doesn't mean you've got more integrity than someone else.

Learning to write as a scientist just evolves, doesn't it? We weren't given much opportunity to write as undergraduates, or at high school. One of my professors when I was postdoc-ing was very picky about style and taught me to think more about it. He's the one I remember. He maybe just critiqued, but explained his critique. So I just evolved as a writer. I wasn't rationally thinking it all through.

But there are things we could do better about writing. I think really what we need to teach students is that it matters. They can find their own voice. They can find their own style. But they should be thinking about it. It isn't something that necessarily just happens or isn't important. It's the whole thing of valuing communication and we don't do that enough.

I enjoy writing. What surprises me is the fact that people read my writing. It started off as maybe even self-indulgent, but now I feel that people value it, all kinds of different people. I'm very conscious of the fact that I have very different people who read what I write. It's wonderful—incredibly good for the ego! But that also becomes a burden. People's expectations—just because you once wrote a nice piece, people expect them all to be that sort of way. It can become quite heavy.

RICHARD ROSE

Richard has an unusual place within his country's psyche: he is a nationally celebrated scientist. Throughout his career as an internationally significant physicist, he has developed an influential national public profile including a regular slot on a national radio programme, as well as writing a recent popular book exploring the potential of science for promoting economic development. During our afternoon together, he took complete control of the interview, waving aside my questions, and engaging in his own way on questions that truly mattered to him: why we must teach our science students to communicate, the public misconception of science (and why it matters), the centrality of the story. After a brief tussle to regain control, I relaxed, sat back, and enjoyed listening to his voice and his forceful views, occasionally throwing in one of my questions when an opportunity arose.

WE ARE TELLING A STORY THAT MEANS SOMETHING

I wouldn't want to think that only older scientists are the voice of science, that they—we—are the only ones communicating science to the public. In fact, I think my passion around science education is to change that completely. I would like graduate students, now, to see science communication as an essential part of their youthful role as scientist. Einstein's best work was done when he was a boy—not when he was an old man with crazy hair. You see this beautiful picture of this young man, in his twenties, and that was the Albert Einstein who did the work that we remember, that changed everything. So, the face of science is often, I think, perverted by the fact that you see these older people out there who are the famous scientists, and people have this vision of the scientist that way. And in fact science is a youthful game.

So, what we need to do is to change that dynamic, so that science communication becomes an essential part of the training of a scientist. And with our institution we've taken the view that we see our role primarily as educational. So, for example, every year our students get together and we get them to give talks, and their talks can consist of, for example, "what are the ethical dimensions of your work?" or "How would you explain this to your grandmother?" They have a choice, they can do one or the other, or "How would you commercialise that work?"

And the reason why communication is so vital is that science has to project its value to the society at large, it has to be of value. It has to engage with the public, who have legitimate concerns about the use of science, and legitimate anxieties about ethical issues. I mean, this thing about these eight babies, it's a disaster for science! I mean, it's terrible! These people are complete hoons. Cowboys! [The Suleman octuplets were conceived in 2009 by IVF.]

So, you know, we do have these problems. And it's no longer good enough just to be smart at what you do, and to publish papers. That is gone as far as I'm concerned. You look at the smart organisations around the world, the best universities, and the on-the-ball organisations, they know this. So I think the model of the young ones learning only to write and speak to their discipline, and the almost-retired scientists speaking to a broader audience, is an old model that might have been true once. I don't think that the new approach to the way we're looking at graduate education is like that. We try to make sure that all the students in the institute get practice at talking about this, we encourage them to be science communicators.

And yet there are undergraduates who think that they shouldn't have to write. They've spent their high school years avoiding writing courses—well they're stupid! And they'll be failures. Look, the fundamental discriminator between those who are successful in science and professionally is their ability to write. That is the fundamental difference. I have students who are very good scientists who can't write prose; they will never become leaders in science. They can't do it. Why is that? Because first and foremost, science is about communicating clearly.

And the most difficult audience to communicate to is your own fellow scientists because the ideas in physics (or whatever field it is) are quite complex ideas, and it's difficult even to explain them to our colleagues. I go to conferences and a lot of the guys I can't understand! Those that I do understand, I remember. And I look to those scientists, and those are the scientists who are successful.

So, whether it's oral communication or the ability to write a beautiful paper that's really crystal clear, the ability to engage in the craft of the science itself quite apart from the public aspect of it—it's absolutely vital.

The other thing is you have to write a lot. You hear scientists complain "oh, I have to spend all my time writing research grants." What I say to such people is "write faster! What's wrong? Why does it take you a month to write a grant?" It takes me a day, or two. Because I write fast. And if you write fast, you're efficient.

I think it's probably true in most professions. In business you have to write clearly, reports and so forth. Convince other people that this is the idea we want to go with. So I think these skills of communication, and in writing particularly, are absolutely crucial. And the writing is the most powerful expression of this because this is where the formality of the ideas can be presented. I mean, you can sway people with an oral presentation, but the evidence base—the argument—is presented in writing and the more crystal clear the prose, the more effective the communication and the more effective you are in any profession. I can pick out the students who are going to be successes. They're the ones who can do a draft of a paper no problem. And they tend more to be female than male, and that's one of the reasons why women are really changing the scene of science tremendously too. As these communication skills become more important in science, become more recognised, you're seeing women pick up a bigger role.

But I say to students "it's not enough just to be good at what you're doing. Why, why are you doing this? How many hip replacements does it cost to do your science?" And I actually talked about this yesterday at this conference. I said "this is how much money I've had in research grants over the years. That's 600 hip replacements, or 120 septum treatments for one year for breast cancer. That's what my scientific research has cost the tax payer. How do I justify that?" Now this has made some of my colleagues uncomfortable, but we must start asking these questions and thinking in that context of science. I often say "Who pays for what we do?" It's people who clean buildings at three o'clock in the morning, for minimum wage. Their taxes pay for what we do. How could we go to that person and say "this is the lifestyle I have as a scientist, I travel to conferences, I do this work, I do this erudite stuff that gets published in journals." How do we say to that person that their money was well spent?

Now, people didn't talk that way a couple of generations ago. There was this kind of a sense of privilege, of priesthood, a kind of a sense of almost entitlement. It's all gone. We don't live in a world like that. We question our medical practitioners, and that's right. So, if scientists don't do that they will not be supported by the public. And, so, it is vital that these young people are able to think about these questions, think about context, think not just about the economic value but about the human consequences, the ethical issues that sometimes arise in science. Also to communicate the beauty and the passion around the subject and get people excited. So they see that science is . . . it's a wonderful thing.

You understand evolution and suddenly the world is just an extraordinary place—you have insights that make you feel more, I don't know, connected with the world . . . that science is actually a beautiful way of looking at things. So, there are many dimensions to this communication thing. But any person, any undergraduate who says "I don't want to learn about communication in science" well they're just fundamentally stupid! I mean, you might as well say "part of the job of the university is to ensure that people like you never get a near a profession!" Of course, you can't say that, but it's the truth! Bright students won't say that, a smart student won't say that.

I've always enjoyed writing. I loved writing at school, and it's always been part of what I do. I even did English literature at university. I did two courses in my first year. I was toying with the idea I might do an arts degree and English was my top subject at school. I loved it! But I had some battles with my poetry tutor in first year. I didn't quite like first-year English and I got into physics. But I've always loved reading, and I read novels and all sorts of weird stuff, and I enjoy that. And I love language and love to read poetry. And most of my scientific colleagues who are leading scientists throughout the world are like that: they are broad-minded people, just like they would be if they were musicians or lawyers. Talented people are interested in stuff. How could you not be? And so, that's not a remarkable thing, it's not peculiar to science, even though it's contrary to the image of science.

People know about the value of technology and science in their lives—antibiotics and so forth—but, there are these other images, and there are these worries that are there as well. Mary Shelley wrote about the fear of technology, the Dark Satanic Mills, which were the result of the industrialisation, way back in the 19th century, so I guess it's not a modern thing. People fear the inhumanity of some of these things, that science doesn't necessarily make our lives better. And there lingers the image of "the scientist alone in the lab," the autistic, dysfunctional person who has no personal relationships and just loves their test tubes, and says "let me get on with it and do my mad thing." Now, that is a falsehood because science is, of all the creative areas, the most social.

You take a writer—they're a very solitary person. A composer, an artist is solitary too. But as a scientist, it is very hard to be solitary. I mean, even if you're a theorist doing your own work you have to be looking at what other people are doing and talking to experimentalists. So science is by and large done in teams, by collaboration. It is very social. If you do a Ph.D. in the humanities, for example, or if you write a book, you are sole author. Whereas with the sciences you more often would have collaborated with your advisors, you'd be co-authors. But this image of the solitary scientist persists, and whose fault is that? Surely it's science's; we're not explaining for people the way it is. So we have a lot of issues there, we haven't put enough effort into that, we haven't been professional enough as a profession to actually explain to people what we do.

I learned to write science when I was doing my Ph.D. I remember the first paper I wrote when I was at Oxford, It just seemed like a natural thing to write a scientific paper that was just like writing anything else. It was just another form of writing. Now I did have a wonderful Ph.D. supervisor who really knew his craft very well; he wrote beautiful scientific prose himself, and I guess I probably learnt a bit from him. But is it different from writing other things—I mean, writing is writing, you know? I don't think there is anything unique about the way you write science, except that you say what you're going to say, you say it, and say what you said. You have to almost exaggerate the clarity that's required. So there is a structure to a paper. But, you are telling a story. I tell my students when they've got to write a paper, "what is the story we are going to tell? Can we tell the storyline?" And I ask them to think, "tell us about it! What is the beginning of the story? How does it develop? And how does it end? Because that is the story we're going to tell in this paper." If we don't know what the story is, you can't write a paper.

You don't sit down and say the experimenter did this, and then this. We had this awful thing at school, you know, "Observation, Results, Experiment." I mean, whoa! Oh my god! You know, the things we do to kids, we teach them this garbage! No, no, you are telling a story and in truth, you've done all these experiments and this didn't work and that didn't work and that didn't work but this did. And we've got to somehow sift out of all this complexity, what we've learned, and throw the extraneous stuff away, and tell a story. If we don't do that, no one's going to read it and it won't make an impact. It's not that we're being dishonest, and if we were it wouldn't accepted in any case. But we are telling a story that means something.

In the top journals, like *Science* or *Nature* or *Physical Review Letters*, they have a process where, at an editorial level before they even send it out to experts reviewers they'll look at the first paragraph of the thing, and they'll say "Is that of interest to a wide audience?" Right, so you've got to know how you write that first paragraph that grabs someone by the throat! That says "my god, I've got to read on here!" If you don't do that, it's not even going to get through first base in those particular journals. Now that's something you have to teach the graduate students. It's not something that is natural, because it is peculiar to the situation. A bit like the journalist, they'll start with something that is right at the top, that's the most vital thing, and it gets less important as you go down. In science, there are ways of writing papers for certain journals, and ways of dealing with a referee's criticisms, and those are some of the peculiar things about writing in science. It's a craft that experienced people like me know, we know the game and can teach to our students.

I think the other thing is to pick out what's important and what are details that don't matter. So in some ways a short paper is a letter. What are the key elements of this, what is it you want to say? And for me, I always say "What are the pictures we are going to show?" Start with "what's the story" and "what are the pictures we're going to put in there?" Because, when most people read papers the first thing they see are the figures or the pictures. And, it's like—well, I grew up on cartoons, comics, and now kids watch TV. As human beings, we are visual beings. The very first thing we notice are the figures or their captions underneath. So I say, "Really, in reading a scientific paper, you should be able to look at the figures and the captions and you should just about get it from that." And furthermore, the figures will guide the process of telling the story. They will be the reader's anchor points for the meaning of the story. So I say "Ok, we're going to write a paper, it's going to be for *Physical Review Letters*, it's four pages long, so there will be four or five figures, little things that fit into a column. What are those figures going to be? What is our story going to be?" And once we've settled that, then all these other things seem to follow.

Figures are very information rich. And of course they are the presentation of the scientific results, the proof. You had a theory, you've got some data . . . now, does it agree with the data? That's the killer bee thing. The words are just guiding the storyline. But the killer bee punch line? In physics anyway it is the pictures. The role for visualisation of graphics is very important and it's always been that way in physics. You know, Leonardo Da Vinci's beautiful drawings—so important.

The mathematics of physics is very important too. What's the mathematic story that we've got here? How can we get this at its most sparing but clear way to write down this without too much detail? Where might we take some mathematical detail and put it in an appendix, so as not to distract from the story? So there are some peculiar things to writing a scientific paper in physics which are different from, obviously, writing a review.

Is scientific writing persuasive? You have to be very careful with this. Something can be persuasive because the evidence so compels you. That has to be the case in science. In one sense, the data, the evidence have to be persuasive in themselves. In that sense, even a person who is a poor communicator, who writes badly, should be able to shake science at the foundations because of the profundity of their discoveries. So the medium is not the message and it must not be so.

So there are people who are very successful in science who are not, frankly, good communicators, not even good writers but their work is so good it shines through anyway. And that's as it should be. They might need a bit of help sometimes from colleagues or editors or whatever. And I defend those people: science needs them, everyone needs them. They have to be exceptionally brilliant in order to succeed; it's harder for them if they're not communicating well. I mean, I say that all scientists who are successful are good communicators. That has to be tempered by the fact that some people are just so exceptional they manage even though they are not. But it is much harder.

I'd say the science itself has to be persuasive, not the writing. However, it does help if the writing is clear. I don't think persuasion, in the sense of an emotion that might come from reading the prose, should have a role in science. One shouldn't be swept along by the beauty of the prose; it's got to be right.

I am a pretty mediocre scientist. I think, to be honest, a lot of my success and my recognition in science, is simply because I do know how to go to an international audience, I can speak English in a way they can understand me, and I can present in a way, thinking about my audience, that they actually come away and say, "you know, I understood that talk! And, I didn't understand those other ones." And so they remember what I did. And honestly, I've had a lot of generosity towards me which is undue, simply because I've made an effort to try to explain.

I think learning to be persuasive is part of the life of being at a university teacher; I often say that university scientists should have a big advantage over those are pure research scientists, because they have to teach undergraduate classes. And there is nothing more humbling than standing in front of a bunch of kids who are saying "frankly I've got to pass this exam, are you able to get me interested in this subject? Can you motivate me?"

I've taught freshman physics all my life. I teach freshman physics here, and the great thing about teaching undergraduates is that they are fearless, in that they're not afraid to ask a dumb question because they want to pass, and they've got to know. So if you stand in front of an audience of your peers, they're not going to ask a dumb question, because they're afraid to look stupid. They'll ask something to show how clever they are. But students ask, "Can you explain what you said?" And so you learn, when you're teaching, particularly freshmen, some of whom are not very motivated, to make it clear. And then when you're teaching students who have chosen to study physics, they're struggling with the concepts and so will very quickly sense if you don't have a deep understanding of the subject yourself. So the clarity and the understanding is kind of a part of what you have to do as a university teacher. My observation is that people who come from universities tend to be better at this than people who don't, though it's not universal of course.

Much of my writing now is first drafted by or with someone else. All of the team will create the story. Someone has to start with a draft. What I will do, particularly with my graduate students, say a first-year Ph.D. student, is we have our first paper, but they have never written a paper before. So we start and we say "now, look, what is the story? What are the pictures and so on?" The most difficult part to write is the introduction of a paper, because the introduction is kind of "what are we going to say? What is the context? Why are we doing this?" They find that the hardest. They're very good when it comes to "what did I do for my experiment" and to be brutally frank, I say to them "why don't I write an introduction? Because I'll give you an idea, you'll get the sense of the way this works." So, the way I'll work in a team is I'll say . . . "we agree on the story" (that's a discussion, right . . .) but when the writing actually starts I'll say "look, here's an introduction. I want you now to go away and write the rest of the paper." So, they will start, and the next thing will tend to be what was the experimental method, what were the results, and so forth. Then we'll start getting the more difficult stuff about the interpretation of that story, and how we would end it off. In the process, it will go backwards and forwards. We don't sit down and write together. I write something, they add something on, I will correct that or make suggestions, sit and talk with them, they'll have another go. We'll go backwards and forwards.

Someone's got to start it off. And you just iterate the draft around. I do have exceptional students who sometimes say "look, I'd like to write the draft" and I've had some who have written a draft and I say "Perfect. Just perfect. Just do it like that." But they're rare.

I've had many students who are almost illiterate (they're illiterate in the sense that they don't write good prose). But the remarkable thing I have discovered is that, in the process of writing a Ph.D. thesis, with all the pain that surrounds that, and me tearing into their prose and rewriting it, by the time they're at the later chapters, it's not too bad. It's amazing! Clearly they just haven't had the practice until that point. And the very practice of writing the Ph.D. is in itself a major learning exercise for many. You have to wonder what they have ever done in the past.

I've had students to whom I say "go and read some novels" and they say "what?" and I say "Go and read some novels!" Don't get them reading science books, go and just read—I don't care what it is, just read something. And, you know, if they're not readers, it's going to be hard, isn't it? I mean, fundamentally the ability to write comes from the fact we've read. There's a resonance to the language. We're not thinking in terms of grammar in a conscious way (even though we might have learned grammar), we write almost instinctively because there's a register of voice that we're used to and we've picked it up from our reading. Things unconsciously become part of the means in our brain and they end up on the page. So if people are not readers, it's hard for them to do that.

The cannon of science, everything that we have, which is standing on the shoulders of those who have gone before, is recorded in writing. If I want to get to know Albert Einstein, I go and read that paper, *Brownian Motion*. He wrote that paper in 1905. And it is absolutely magical. I can only read it in the English translation. And I can see right inside the mind of that man, it's like I know him as a friend from that record that's left. It's nothing to do with the pictures that have been in *Time* magazine or any movie of Albert Einstein riding a bicycle that tells me anything about the man. It's what written, that's the record that remains. And that is all we have in the long run, in science, is the written record of what we do. We all die. I'm not going to be standing up in front of an audience in twenty years' time, but I would like to think that someone might read one of my papers and think "that's really interesting."

There is the sense in which the recording of, the visual through digital recording medium, means there is another mechanism for people's science impact living on, and you can see Richard Feynman in these wonderful lectures he gave in the early 2000s. And it's fantastic to watch but, nonetheless, you're really not quite in the mind of the man at the deepest sense unless you go and have a look at his scientific writings. Because that was where he was writing for that audience, his scientific peers. It really mattered for him. So that's the ultimate record that we have. I think that's surely the most important thing. That's what we leave behind.

CAMERON

Cameron, a close colleague and co-author of Richard's, is also committed to changing the public perception of science and encouraging government to recognise the potential of science in relation to economic development. In mid-career, he made a choice to engage in outreach through a significant blog, and thus has an interesting perspective on the role of social media in science, as well as a perspective on writing in different genres. He has a warm, self-effacing manner, and his style of speaking entirely lacks Richard's forcefulness. But this in no way diminishes his influence and perceptive vision for how writing and communication will influence not just the future of science, but also the future of government and humanity.

IT COMES WITH A KIND OF MATURITY, KNOWING WHAT WAY YOU CAN PUSH THE BOUNDARIES

My blog evolved out of my role as Deputy Director of this institute. I took on that role in 2008, and one of the things that the institute has always done is that it's had a big emphasis on outreach. When I came on as Deputy Director, our Director was less into the outreach and so I decided outreach was one of the things I should take up as part of that leadership team. I thought about what I could do, and I'd had a little bit to do with the science media centre, and I was aware that they were planning to start a blogging site, and so I just decided one day that that would be a really good way to start getting into the outreach. And it was something that was a comfortable way to get into outreach, although I still remember I was fairly nervous when the blog went live, but still it was an easier way into outreach than some of the other things I could have done, like going on the radio or trying to do more active things. It was something I could do in my spare time and just work on and practice. So that was my entry way in. I guess I've been at it for about four years.

Actually I have to say my blogging's dropped off—I'm only putting out probably at the most a post every two months at the moment—because I've now written a book and I have a column, so my writing's spread out and my blogging has dropped off a bit. I guess at the peak of the blogging, when I was putting out things regularly, I was probably getting a readership of about 2–3,000 people per month, something like that.

I was blogging about innovation at a time when there weren't many people writing about that area, particularly from a scientist's point of view. I've had several people point out that I fortuitously filled a market, a niche. And then also I guess there are a few topical things that happened that enhanced my readership. So, during the Fukushima earthquake with the nuclear disaster, I wrote some pieces which got picked up by a national newspaper and that generated a lot more traffic to the blog.

Right at the outset I had a goal. Richard and I talked about some of the things I might do, and he said one of the things that he'd been able to do, and that he wanted the institute to continue to do, was to offer thought leadership for the country. So that's what I've tried to do. I've tried to be analytical and think quite deeply about some of the challenges facing science in the country, and also—from my blog at least, in quite a few of my articles—to aim it at policy people. So I was trying to push topics that I thought would be of interest to them, and I knew I was reaching chief executives in ministries and politicians, so I knew that was successful. I had a clear strategy; I had a whole lot of stuff that I wanted to cover and it was a matter of just putting some thought into particular topics.

The other thing that happened is that—and this was interesting—having signalled my willingness to engage through blogging, I then started getting a whole lot of other things happening. So I would get requests to talk and I also got a slot on radio, and so then if I was going to do a particular interview on a topic I would try to write a blog post—I had to prepare material anyway. So I had this big, overarching theme to my writing, and then, if I saw something topical I could write about, I would jump in and do that as well.

My overarching theme is about how science contributes to the economy, and also what is it that makes a successful science system, or what makes some parts of the world more innovative than others. And I actually turned this into a bit of a research area, and so sometimes I was writing about original research that I'd done on this topic and other times I was drawing on work done by other people.

As well as this, I have a monthly column in a business magazine, and I've also had op-eds reasonably regularly. So there has been a shift in my writing—this does tend to happen in a physics career. I mean not the other side of writing, popular writing, but as you move up in physics you tend to become the leader and, because you're working with students, part of your job is to train them to write so they become the primary author on their papers. That's generally true. I guess the other part is that traditionally you end up writing grants; the group leader writes the grants and the students and post-docs write the papers is how it's traditionally done. I still have to do that, but then I also do a lot of this other writing now.

I enjoy writing. I mean, I have to say having a deadline on the column, that's hard work. It's not that writing is not work, but I do enjoy it and I particularly enjoy the popular writing—it's been a bit of an experiment for me as well and I do like doing new things. I really enjoyed writing the book—that was a really good experience. It was really hard work but good.

I was writing the book with Richard, and it was probably not a typical experience because he passed away halfway through. So that was a challenge. But then again, you know, we had quite a bit of momentum, so even though we still had about half the material to write, I felt like there was enough momentum. What I did find hard was . . . the bits that I knew he was going to write, I found it hard to write in his voice, if that makes sense? No one's yet been able to guess which bits Richard wrote and which bits I wrote, so the style's kind of consistent. But I found it uncomfortable and awkward—trying to put words into Richard's mouth, I guess, is how I would put it. That was a strange experience.

I think I'm probably a drier writer than Richard, less flowery. And I think maybe it's blogging that has changed my writing; I've had a lot of practice and I've made a deliberate effort to simplify my sentence structure when I blog. Richard's writing in a much more complex way than I am. It's very clear when

Richard writes, but it is more complicated. I do find myself to be a very functional writer—that's how I'd describe my writing. You know, I'm not sure I'd read me.

Richard has certainly influenced my interest in writing. When we planned out the book, we planned out the narrative and then we both went away and wrote. For the first couple of chapters, he was in Cambridge, and I was travelling as well, and we both came back and compared notes. And actually we'd both written stuff that went at the same pace and I wonder if, having read lots of Richard's stuff, I was anticipating how he was going to approach it, and then I think I probably tried to match the tone I was expecting. Obviously we've done a lot of editing since then, but those first three chapters really did feel like they went at the same pace in the way we skipped over the ideas.

Perhaps the difference with my popular writing and my writing for a scientific audience is that, when I'm writing for a scientific audience, it's typically a collaboration. So, even when I'm writing a grant, I might be the lead person but I won't necessarily have the expertise to write enough detail on a particular topic. So I think one of the things that's different is that you know it's a collaborative process, writing a grant, and so you do have to have a discussion to start with. But often you need to start, so I will need to start by putting some text down just to get other people thinking along the same lines. So I might have to write something that's very structural, that has the structure of what I think the grant should look like, and then I've got to put it to my collaborators.

Then I really start thinking about how I want to explain the ideas; and so we might get some structure to it; we're going to give the general introduction to the idea, get into some specifics, say why we're the people right, etc. Then you really actually need to put some meat on those bones and think about what you need; you might need to think about some examples, or some metaphors. You need to consider how to spark people's interest in the grant. So you've got to have a punchy interesting introduction. You'd perhaps go through that process where you're trying to write everything, make it interesting, and then often you start losing accuracy and, of course, you've got space constraints with the grant proposal. Then it's an iterative process as you try to keep the interest in the proposal while making sure it's accurate, but also keeping to space constraints.

The grant goes to a physics panel—well it's broader than just physics, so it might be physics and chemistry. It's a relatively general audience; it can be very challenging for someone in chemistry to understand a proposal in theoretical physics. It's reasonably high level. So you're kind of negotiating between wanting to make it readable for a more general audience but keeping it specific enough for somebody who is on the panel who wanted to make sure you could get the detail right.

I do edit as I go, and so I might get halfway through and then go back and re-edit because I'm not happy with how things are going. And then I like to put it out to my collaborators to have a look at, get their comments back, maybe get some edits from them. So having gone through some editing and writing, I then like to have a little bit of space; I often like to have a couple of days just to come back and look at it fresh and then I'll almost inevitably be quite dissatisfied with it.

For a paper, thinking about the journal we're targeting is almost the first conversation. You need to know at the outset what type of journal you're aiming for. And so for the more prestigious journals you're going to go for a quite short succinct paper or for the less prestigious but more technically specialised journals you'd go for a longer format that's more in depth. So you need to make a decision right away. It doesn't mean you'll necessarily get into those journals, of course.

I definitely am thinking of my audience. I don't always. I don't always feel a difference between, you know, when I'm writing different things. I mean I do think about it, but it's often the structure that will change rather than my style of writing specifically. It's difficult to describe. I guess it ought to feel different when I'm writing for different audiences but it doesn't always. So it is something that puzzles me a little bit. I suppose I might be more playful with the broader, more public audience—but I can also be quite playful in grant applications too, to try and capture people.

When I started blogging I was very uncomfortable because, as a scientist, you try and write without opinion—do you know what I mean? You know you're writing based on things you're very, very certain about and there might be a little bit of speculation at the end of your paper but it's clear you're speculating. But when I started blogging, I had to put opinion in because I could see from the stats on my blog that the more opinionated my blog the more readers I got. And I think, for a scientist, when I started out doing that, I was quite uncomfortable and I was particularly worried about criticism from my peers. The other thing that happens, of course, with science is, if there's a stupid mistake in your paper that will get picked up through closed peer review, you won't be humiliated in front of all your colleagues. But when you're blogging or when you're writing popularly there's not that private checking necessarily going on, and I think I was quite nervous about that when I started, but it's something I've gotten used to.

When I'm writing, I do think I'm being persuasive. I think so. Although I said I am putting opinions out there, I'm trying to make them evidenced-based opinions and so probably one of the things in my writing is I do draw on evidence a lot and try and explain where that evidence comes from. And in scientific writing, that's even more evidence-based.

Scientific writing is probably easiest for me. Probably just because I'm much more practiced at it and it's much more based on my own work. It's the stuff I know best; the further I go from where I'm grounded in my own work probably the harder it gets. But, I'm a theoretical physicist, so compared to solving difficult equations in physics, writing for the blog is not something I think of as difficult. I guess it's time-consuming, it does take work, and I do need to put effort in, and I need to motivate myself to do it, but I wouldn't say I find it difficult.

Finding the time to write is actually really difficult, that's the tricky thing. And especially what I've found with writing the book is I needed to take out big chunks of time to do it; you know with the blogging I was able to write something on a Sunday night or on the plane or something like that. When it came to the book, I needed to clear a week, and once I'd spent a couple of days preparing my material and thinking through a bit about the structure and stuff, then I would actually write quite fluidly. So by the end of the week I was writing a lot faster and a lot more fluidly than I would at the start—so actually finding that time is critical and, of course, when you're trying to sandwich it in with your academic job and teaching and things, that's quite hard.

I'm a bit of an idealist I suppose, and I do think that it's important that scientists do communicate with the public. You get all sorts of positive feedback for doing it as well, so that's always good. When you see that you're having an impact and having an effect and changing people's opinions—that's really positive. And I want to make a difference I guess. You also get direct rewards for doing it—I learn a lot as I write, and by reaching out to new audiences I meet new people and I learn from those people. In my research group we've picked up funding because someone's read a piece I've written—actually understood what it is we do, and come to us to say "well, could you do this for us as well?" Or "I've got this idea." But fundamentally it comes down to wanting to change the way the country does things and the way we think about science. The writing is a way to achieve that. I've found it's really effective.

There might be times when you just go away and do the science and then finally report the results. But these days communicating with your team is very much part of that, part of the scientific process, because science is so much a team effort these days. And if I've got a student or a post-doc in my group, one of the things I'll get them to do is write as they go because (a) it helps keep their thoughts organised, but also (b) it's a good way of communicating to the rest of the team as to what's going on and what they're thinking about. So I do think writing is now embedded as part of the scientific process.

When you read equations, they should read like a good sentence and I don't think everybody gets that, but that's certainly how I feel. Equations are part of the language and they should fit into good writing. I think there are still corners of science where you can beaver away just writing equations perhaps, some areas of mathematics—but most of science (and this includes almost all of physics) you've got to be able to communicate, and writing would be part of that. For me, equations are like writing. Or you could look at it the other way—sentences are equations. That may explain why my writing is so functional! I don't know.

To me sentences and equations are not very different. And I think again perhaps, if you work in pure mathematics—that might be the case that those equations don't correspond in sentences. But in physics, these equations describe part of the world, so they're descriptive. I do have the ability to read equations that don't necessarily have a physical meaning. But generally I'm not comfortable with an equation until I can put a sentence structure to it or a physical interpretation. That's a very wordy way of explaining what I mean!

I think people are communicating much more broadly at all stages of their careers now. There are far more outlets now for scientific writing than perhaps there were 10–20 years ago. And I know we have Ph.D. students here who blog and who tweet and who do all sorts of things—which wouldn't have been possible in the past, when I was a Ph.D. student. Writing's become much more democratised in the sense that we can self-publish through our blogs. I think people have embraced that and we're maybe not following the traditional model any more.

You know, having said there's a whole lot more outlets for writing these days, it can also be very distracting. Where should you put your efforts? You do have to mentor people, and so you have to talk about their priorities and that they do need to put time into things. There are people who think that I've probably hived off and starting doing this stuff too early as well. There have certainly been colleagues who think that in your early 40s is when you're at your peak of your scientific productivity, and by diluting that I've probably hurt my academic career. There's no doubt that the popular writing I do takes away from my scientific productivity; but on the other hand I think I'm doing different work because I'm doing this writing as well, and I think I'm probably doing more interesting work because of the wider exposure I've got with a wider variety of people.

I see the other point of view, but it also worries me. What are you missing out on when you're one of only eight people in your field? If there's a circle of you all talking to each other, there are opportunities you might be missing for taking your work in new interesting directions. There's a risk, a risk that people's work just disappears down rabbit holes—and this is one of the things that comes out when you look at how innovation and creativity work. The social network that you're operating under is crucial for creativity and innovation. So talking to people about your work is important and not just those people that are in your own field.

There are conventions for writing in my discipline, but some of the most pleasant scientific articles break that mould. Certainly those types of conventions are a good rule for people starting out, and of course a lot of physics is written by people for whom English is a second language and so if they're going to write effectively they'd need some tight constraints. So there do tend to be constraints, passive voice and so forth, but they're not always followed these days. I wouldn't always follow them. It comes with a kind of maturity, knowing what way you can push the boundaries.

CHAPTER 3
THE RELUCTANT WRITERS

> No, I don't enjoy writing. I would prefer to just upload the results, upload spreadsheets to come cloud somewhere and that's it; I can get onto the next project. I don't think of the projects I'm doing in terms of paragraphs . . . so having to translate a project into words like this is . . . I would prefer not to have to do that.
> — Emerging scientist, Cognitive Psychology

Many of the participants in this study revealed positive attitudes about writing (see Chapter 7). There were, however, some scientists—just over 11% of the total sample—many of whom were successful writers, who disliked writing, or who saw writing as largely irrelevant to their science. Two things were striking about this group of reluctant writers. First, they were far more likely to fit Holyoak's definition of the routine expert, writing for a narrow range of journals and peers, and not venturing into more diverse genres. And secondly, most had found a way of compensating for their weaknesses as writers, the most common of which was relying on co-authors.

One participant, a senior scientist in the medical sciences, described her problem as being a perfectionism that made drafting a challenge:

> I'm a very good writer. I hate it. I hate it. Well, really, I'm a very good editor and a lousy writer. I have this terrible problem with the tyranny of the blank page. You know, I sit down and look blankly at it. I find it difficult to get started. Which is why I'm a much better editor. Somebody else has started it. There's something there to work on. It's the getting started that I find difficult.
>
> I think the reason I don't like it and prefer somebody else to get started is that because I like it to be right—I'm a bit of a perfectionist. If I'm writing from scratch myself, I waste a lot of time getting it word perfect; just getting it down and going back and editing it. I'm a slow primary writer. I edit my own work as I'm writing. (Senior Scientist, Neonatology)

Her primary solution was to always work with graduate students or postdocs—once she had a draft, she was confident in her role. But she had also developed a range of approaches to kick-start her writing and bypass her perfectionism, including dictating her work or a technique that she called "the three-hour exam" where she forced herself to prepare for and then write to a goal for an extended, but defined, period.

Those who disliked writing—as opposed to those who simply struggled as writers—tended to demonstrate different beliefs about writing compared with those who actively enjoyed writing. They were more likely to ascribe a narrow function to written language, to see figures and equations as intrinsically more complex and more nuanced than words, to place less emphasis on audience and story, and to define a narrow role for themselves as scientists. They tended to be inward rather than outwardly focused, defining a narrow community with which to engage, which in some instances seemed almost to become invisible during the writing process. "I write for myself, because I am writing for people exactly like me" (Emerging Scientist, Chemistry).

I have included one short and three extended narratives in this chapter which illustrate the range of the attitudes and beliefs of those participants in this study who disliked or struggled with writing. The first is Wendy, a doctoral student who loves to write, has experienced joy and confidence in writing but is struggling to come to terms with unexpected, immobilising criticism. Next, Jane attributes her problems as a writer to her inability to find a writing/research community. Her story of extensive isolation was relatively unusual amongst the people I interviewed (only three interviewees discussed this problem: all were women and all were emerging scientists), and I include it as a comparison to the many other voices in this book that focus so strongly on community. The third narrative tells the story of a highly successful mathematician who revealed a striking struggle to write but who, with a wide range of support from parents, teachers, advisors and co-authors, has become, not the plumber he might have been, but a highly regarded writer in his field. The final narrative comes from a scientist who sees writing as tangential, if not irrelevant, to science. My interaction with him might be described as a clash of ideologies, but he certainly voices an emphatic perspective.

WENDY

Wendy is a mature doctoral student working in an area of women's health. When I interview her, she's in the second year of her research, and she's struggling. She faces numerous obstacles: a lack of funding, no research team around her, and—most problematic of all—an advisor who she perceives to be erratic and volatile. I interviewed her over two sessions—we stopped the first session because she was so distressed about her situation that I switched off the recorder.

What puzzles her most is the critique of her writing: she was a confident writer through school and through her undergraduate and first post graduate degree. She has had a successful career that required her to write about women's health for public audiences—and she's passionate about communicating her work. In this short section of her narrative she links her problems with writing to her philosophy of life. Since this interview she has found a new advisor and successfully defended her dissertation.

MOST OF SCIENCE IS ABOUT PEOPLE

I always think you have to communicate well if you have gone to the trouble of actually doing something. You've got to be able to present that information well for whichever audience it is. So if you are writing for a journal it's presenting it one way, but importantly for me, because I am quite a people person, I like to see that it's translated into readable information for the people it concerns. And most of science is about people.

One of the ladies in my research actually said that me that she had been asked, because she had a particular health condition, all through her life by several major hospitals to participate in research and she never wanted to. She said "I've always hidden, because I think people get used." But for some reason she connected with me, and she said to me at the end that she really enjoyed being part of the study, and she had an understanding that I would take responsibility for communicating the results in a readable or understandable way for lay people. So I took that as quite a challenge and I understood the responsibility involved in that. I think that's important.

Until last year I may have thought that I was a reasonably good writer. I now think, as a result of what my advisor is telling me, that possibly I am a very poor writer. I'm not sure whether I am misreading or it's my advisor's attitude, but really there's not a huge variation in what I see other people have written and what I see I have written myself. And then what comes back to me, what's fed back to me by my advisor, is that my writing is not good. I can't quite see it because she is saying general terms like, "you have got to learn to critically evaluate." I understand what critical evaluation is and I think I am doing it, but obviously not to the level she wants. And I don't understand how I can see this in somebody else's writing but not in my own. So I have been reading other theses and, while I can see some very well-written ones, I have still got that gap in my own area, so I am thinking I might go and find some help. The worst part of missing something is not knowing what you are missing. If you knew what you were missing, you could aim for a target. This is my biggest hurdle as a writer of science: I can't really get much past my advisor.

Chapter 3

When I was at school I always knew I was a reasonable writer. My English teacher loved everything I submitted and that gave me a lot of confidence. It was mainly critical analysis of literature and I really enjoyed it. I remember the feeling of writing and being absolutely emptied out after sitting, perhaps on a Sunday afternoon, writing the assignment and getting so involved in it and how to write it well, and that feeling of being absolutely emptied—it was so wonderful. I was drained physically and probably emotionally, but I had a wonderful feeling of lightness. And I'd submit it and it would come back invariably with A+, and so I did get the English award at the end of the final year, and I have recalled this over these past 18 months when I felt that my writing has come under such close scrutiny. You think, "well, I don't know. I used to write well." I've never had the criticism I've generated in the last 18 months for my writing.

Writing's a process for me, and I just rewrite and rewrite and rewrite. And that's probably got me a little bit into trouble with my lit review because I write a draft that would be where I'm at, and I would be expecting to rewrite it and rewrite it and rewrite it like a series of layers unfolding. This is how I view life; I see life as an unlayering. A peeling back until you get to the essence of something. It's not "oh, I'm this person now, I want to be that person—so I'll do this this and this to get to that point." I don't see life like that. And perhaps that's been unfortunate with science because you are meant to have, according to my advisor, this body of work that's very clear and coherent and explains exactly what you intend to do at the beginning before you go in and do it. And she's probably right in that regard. But I would have written the lit review and thought "oh yes, I'll change that and I'll change that and I'll change that." So that is how I tend to write and think and then change and think. It's great having the computer so you can do that.

Possibly it's a good seven or eight year process to get to grips with scientific writing. I mean I did write science when I did my undergraduate degree, and then I did a post-graduate diploma, and we had a short dissertation to do, and I wrote then and I didn't appear to have any problem with that. When I did my master's I didn't have too much issue with the thesis I wrote but, yes, that all seems to have come a cropper at this level.

In terms of style, there seem to be variations which I am still getting my head around. Like I was reading in somebody's paper, and I have seen it in some theses, about when you have animals that you are using for experiments and they have to be killed. We used to say "the animals were killed," and now they are saying "the animals are being sacrificed." I thought when I first read it that must be the correct term, but I recently read a book which a colleague lent me when she and I were talking about the difficulty with writing, and in this book the author mentioned this word "sacrifice" and he said "they were killed—let's stop getting pretentious about this." It's quite a religious connotation, and when I first read it I had the feeling they were sacrificed like they were on a pyre, and there was incense and a fire going—but no!

And it's almost like you come up with new words to give more impact. There's always something new coming out and you've got to keep yourself abreast of it, but whether it's any better or worse than what's been written before . . . ? I prefer something that's fairly clean and not too wordy. And I like the statistics to be readable and I am very good at picking up other people's mistakes in a paper, grammar mistakes or a word missing. Just a typo error always jumps out at me. But apart from that I don't think I could say very much more about scientific style because my confidence isn't really great with writing. I feel that I am probably not the best person to be talking about that.

I wrote from the earliest age. I can remember that I could write before I went to school and I used to love writing on projects at elementary school. I think, for all the Catholic schooling that you could disagree with, the one thing those nuns did well was they gave us an exceptional grasp of basic English grammar. And in fact, one of my friends in Australia is also writing a Ph.D. at the moment, she's a sociologist, and she said something similar when I spoke to her end of last year. You were very conscious that you were taught to write well and it was very disciplined. There was always lots of writing.

For my diploma we had a health communication course, and I think for the other courses I averaged an A- or B+, but that one we all got Cs or B-minuses. We had to write essays about communicating health messages and none of us actually knew what we were to write. The first one we did we had someone teaching the distance course who didn't communicate to us what was required at that level, or what he was after. We all wrote essays virtually off the top of our head—very creative essays about what we thought about communicating health messages. It came home to me in that course that no one in science was interested in the creative side. They may be interested later on when you have suddenly got to pull it out of a hat and come up with something creative, but at that point in time it was squashed. There was some dissention amongst the ranks because the students thought they'd done pretty well thinking up all these new and novel ways of communicating health messages and expressed them well, and they walked out with B-minuses and Cs because they hadn't formed the bulk of the assignment or essay around the existing literature.

So that was a bit of a wakeup call. And I often think, where does the creativity go when it's squashed at an early stage? If you spend years like that we can lose that creativity—but how you incorporate it is another thing.

For my master's, I spent two years part time writing a dissertation. That process was interesting; I didn't find it too onerous. I enjoyed what I did. I enjoyed interviewing the women and relating it to what had been said in the literature. It was an area my advisor, K, didn't know very much about.

Yes, she was my advisor then too. There were obviously some issues there with communication all the way through because I was using grounded theory and she was not familiar with it; she did give me some instructions that I had to follow because I said "well I'm not going to do any quantitative analysis" because with grounded theory you are not meant to. But she told me I had to go out there and get some quantitative data and do the analysis and put it in the thesis. When the thesis came back, I had very favourable comments on what I'd done but both reviewers were grounded theory experts and they said "why is the quantitative material in there? It shouldn't be in there."

My colleague J and I talk about a course in scientific writing; not involving huge amounts of assignments if you are already writing a thesis, but so you can see why something's considered good writing and why something's considered not so good. You would think at this level that you would have that understanding anyway, but when you get feedback and you are not quite sure what the problem is it's hard to go and find help and hard to find where you can analyse what you are doing wrong. But I've this week been over to the learning centre and discovered that they do offer a one-on-one, perhaps an hour a week, with the writing specialist—I wasn't aware of that. And you write something, and then they take some of your writing to critique it.

I wonder how many science writers they have over there because I find when they do run the post-graduate courses here for Ph.D.s or post-graduate students, they tend to be led by people who have written in the arts disciplines. I remember one I went to and a few of us science writers were talking about having to write and produce a paper. Those sessions we had didn't really throw that much more light on it because the people weren't aware of where we were at.

If and when I finish this thesis I often think I'll write poetry. I'm a little bit of a closet poet. I love poetry. I love the way it's only a few words to say something and you leave it to the reader to interpret it. I find it very soothing, and I find that poems I learnt as a child have stayed with me all over the years. It's fall right now, and I'm riding to work and I'm looking around thinking, yes "seasons of mist. . . ." It's just so beautiful and it's amazing how you can learn a poem and it will stay with you for the whole of your life. Now how many things like that happen?

All the words come back in various guises. I mean, I wouldn't have thought of that probably for years but it's always there, like it's been planted, it's a seed. And I think writing can be planted as a seed and then can come out and mature in later life—it's a real blessing.

JANE

Jane's quiet voice described to me the darker side of science's social context: what happens when you're not part of a team? When people you expected to collaborate with leave? When your doctoral advisor doesn't communicate? Jane attributes her lack of confidence as a writer to her social context. Her story was startlingly different to that of most of the people I interviewed, and I'm still puzzled by the causes of her unwanted solitude. Was it because she's a woman? Because she's working in a particular discipline (community ecology, macroecology)? Or was it just a matter of chance? But it is also, I think, worth noting the support she did acquire along the way. It's clear from Jane's story that doing science alone is a difficult task. But is Jane's real problem "a fear of the blank page"?

I Think I'm a Bit Scared of Writing

I came to this university to collaborate with people that I had worked with before, but they'd moved on, and I didn't end up forming good, strong collaborations with anyone here. In the last year or so I've started some collaboration with people here and so I probably will end up writing with people I work with. But over the last six years or so, I've been doing most of my work by myself or with people from overseas, and so that's how I ended up doing a lot of my writing alone.

I almost always come up with my own ideas for research, which again is how I end up doing it by myself. I'll have an idea, and sometimes I'll go and ask people if they want to be involved, and often people will say yes. But they don't really mean it, because everybody's doing way too many things.

I got a grant and I collected some data and added it to some historical data. I ended up writing it up with a guy in the UK. It started when I came here and I was talking to a guy here, and he knew of a data set we could use to test some of the hypotheses that were out in the literature. So I wrote the grant and did a lot of reading and writing at that time.

That was the first big grant I'd ever written, and I had quite a bit of help from two people who were actually on the grant, neither of whom ended up helping with the project at the end of the day. I haven't published with them because they just ended up not really being interested. Once I'd got the money then I wrote a number of smaller grants and got additional money. I didn't really do any writing for a long time—a number of years I guess—while I was collecting and analysing the data. The writing I did do was like progress reports for funding bodies.

I don't tend to write while I'm collecting data. But while I was doing data collection, I was writing up things from stuff I'd done before. I tend to write stuff up after I've collected and analysed, because it takes quite a long time. When I came to analyse it, I started talking to a guy in the UK and he had some good ideas for the analysis of the data, and so I went over there to work with him and I spent two weeks there and we did the analysis and started writing the paper and had quite a solid draft before I even left.

I had read one of his papers and it had a bunch of abundance occupancy modelling ideas in it. His paper actually had a whole lot of mistakes in it, and I was trying to use his equations and they didn't make any sense, and I didn't know if it was my bad algebra or what. So I contacted him and said "hey I can't figure this out" and he said "oh I'm so terribly sorry, I was so proud of this paper and it's got all these mistakes in it." And he was all sad, and I told him what I was doing and he thought it was really cool. And then I asked him how I could use his models and he said "oh well what about doing it this way and that way," and then he was going to visit me. But then he ended up having an unexpected baby and he couldn't come and visit, so I went and visited him the next year.

So it was really a slow process and it stalled at various stages because I guess I wasn't sure how to proceed. There were just so many options, and I tried to work through things and it was really complicated and I was just doing it on my own and so it took really a long time—much longer than it probably should have. But, you know, when you are doing things by yourself. . . . I have a lot of self-doubts, I guess, and a lot of times I just didn't think I knew what I was doing. But in the end, I went over and visited him and that was actually a really positive collaboration. I hadn't had that experience since my post-doc—of really talking with someone who wanted to listen to what I had to say and who thought that what I was doing was interesting. So it was quite refreshing. I'm not sure if it will be an ongoing collaboration. He's really busy and I'm doing other stuff now. I don't know.

It took another year to finish it off because we added some data, and changed a few things around, and so I redid the analysis and then finished off writing it up. That whole process probably took another year to do. It's quite complicated stuff.

The methods were pretty easy to write. I think I probably just wrote those and my collaborator probably wrote some of the modelling bits. And then, for the introduction, we talked about the process of what we were doing and the reasoning behind it and so I got an outline together and then he filled bits in. It was a back and forth process. Then I wrote the discussion when I got back here, so I guess we didn't have a full draft because we didn't have really a discussion when I left. But we had introduction, methods and results which was pretty good for two weeks.

When I'm writing a paper, I have long spaces when I'm not writing, and I don't think that's actually a good way to do it because I get a bit scared of it, and I get scared to come back to it. I don't have a lot of confidence in my writing. So to get that paper done, I had a writing group. We tried to write every day and that was really what got that paper finished off—sitting down and doing a little bit every day and not feeling like I had to do the whole thing in two weeks. Instead, it was just chipping away at it and that was really good. But then once that was finished and I finished another thing, I've stopped again now. And I mostly don't write—I haven't really written anything for months.

I work in quite a large field so there are lots of people and there are people in different parts of that field: so there are theoreticians who come up with ideas, and then there are people who have a real, practical, applied focus, and then there are kind of people who might collect data to test theoretical ideas. I'm one of the people who tends to get real world data to test the theories. I don't come up with theories, I'm more just trying to understand them and see how useful they might be. But I'm also not really a conservation biologist, so I don't tend to write for an applied audience like how you could actually use this to save a species. So I tend to try to write so that all three groups can access the work. And that's quite different to how you might write if you are only writing to theoreticians—it's quite different from how you would write if you were writing for like conservation managers, so I try not to use jargon. I try to write it at the most simple level I can, just because I think it makes it easier for everybody.

I believe in making scientific writing as accessible as possible. I know a lot of people don't believe in that, they believe that what they do is too complicated for the average person. But ultimately it would be nice if people outside your field could pick up a paper and read it.

I tend to try to be quite factual about the way I write, which maybe isn't as flashy as you could be. A lot of people are very persuasive, aren't they, in their scientific writing, but I don't believe we should try to be persuasive. I think we should try to be objective and present alternatives so people can make a decision. But you do have to present the right stuff so they can make that decision, which I guess is kind of persuasive.

I think I am a bit scared of writing. I've had some bad experiences of writing. It probably stems back from my Ph.D. where I had a really funny man for an advisor. He was not very interactive; he didn't really speak, and I never really had a conversation with him—it's not a good position to be in! I had these ideas and I went out and did all this work, and I thought it was alright and I based everything that I did off the literature, and I would email various people in the field and get their advice and help, and I went to a few conferences and talked to people and got ideas off them. But he was really not—I don't know—he just wasn't interested. We were all very scared of him because he would be quite rude if you said something that was sort of stupid, you know, so people were very intimidated by him. As I got further on in my studies, I realised that it was less my fault and more that it was him that was strange, but for a long time I thought "oh there's something funny about me," you know, as you do when somebody acts strangely towards you.

So the process of writing my Ph.D. was very solitary; I mean I did it absolutely by myself. I would give him drafts of my chapters and he would hand it back and there would be nothing on it. On the page, out in the margin, there would be like a cross or a question mark, and I would have to go back to him and say "why is this here?" I would have to go and ask him about every point. So in the end when I needed to submit, I would actually print off a sheet of paper with specific questions for him like, "do you think I should include this in Chapter 1 or Chapter 3?" because I was really struggling and looking for advice. The only other person I had who read my thesis was a master's student there at the time. She was a student of his as well and knew what he was like. She had worked in a similar system (not the same topic) but she knew a little bit about what I was doing and she was the only other person who read my thesis. I felt very uncomfortable during that process.

I had a committee, but I'd had a very bad experience in my oral exam that you had to do a year into your thesis where they say whether or not you are good enough to continue. I found the people on my committee really intimidating. I was really quite intimidated the whole time I did my Ph.D. There was one guy who was on my committee and he was really nice; he liked my stuff and he thought it was cool and he helped me out a bit. But he was really busy. He had 15 Ph.D. students, so I couldn't get help from him.

But I knew, from talking to people, that my stuff was good and interesting. I had some confidence in my ideas, but I just had no help during that writing process. So then at the end when I handed it in and had my final defence, all the people said that it was just an awesome study and they were all like "can I get a copy of your thesis afterwards?" and, you know, blah blah blah how great it was. And I was absolutely bowled over because I had had no feedback, so I didn't know what I had done. That was the last time I saw my advisor, and he just held out this limp hand and said "congratulations." And he was horrible to me in the exam, and I just thought "what?" Anyway, by that time I had realised that it was all him and he was odd and so I just sort of said "thanks, see ya!" and I never saw him again until years later. He is an amazing guy, and he sat in his office under his little lamps and did his thing and wrote books. I think he is a nice person, sort of, but he just had no idea how to be a mentor or how to supervise people.

Then I went into this post-doc where I had three bosses and for writing up I found it really hard because they had three different opinions and so I really struggled in the analysis part of the work, just getting what they all wanted. So that took a really long time and then when it came down to writing I guess I had had this bad experience and so I didn't have a lot of confidence. So I went to them for lots of advice and that meant that I gave up ownership a lot of the time, and I was trying to write to please them rather than sitting back and saying "OK this is my study—what did I do? What's important?" And I think I'm now, eight years later, trying to unlearn a lot of that, but I didn't even realise at the time how that was affecting me. And then my boss was totally hopeless; I could give him a draft and he wouldn't get back to me forever, and one of the papers took about five years to produce because it was just all these holdups and people disagreeing and things getting changed and reanalyzing about six times and no consensus on the writing. And then we sent something in and then the reviewers had all these problems and then we had big discussions about how to deal with those problems, and you know, just round and round and round.

So that's the context, the background to where I've ended up now. And now I find that again I am on my own, and I really haven't published a lot of papers. I only have about 12 in total, and I think I am quite good at writing; I mean if I was honest about it I think I am not bad at writing. I can write well so people can understand and I'm quite good at helping other people improve manuscripts and contributing to manuscripts. But when it comes to a study I've done by myself and sitting down and saying "what have I done?" and writing it up, I just lose all confidence and I have a fear of the blank page and I'll do anything to procrastinate my way out. Mostly what I do is analysis, and I'll just say "oh well, if I just did it this way and that way" and so I spend all my time analysing instead of actually knuckling down and picking a particular analysis and writing it up.

I learnt to write by reading other people's writing. I was lucky because I had quite a good background coming into university, and then in college I enjoyed writing; I liked doing essays; I loved all my undergrad—I had a great time—and I wrote a really good honours dissertation, and I got an awesome mark. So I was coming into my Ph.D. with a lot of confidence and thinking I was a good scientist, and it all just sort of got ripped away. So I think, ultimately, I have got quite a good writing style, but it's been 15 years since I finished my honours year, and I really don't know how much I've progressed since then. It's been a fight for me since then. Which is very disappointing to me.

Another help was, when I was working in an earlier job, I worked with a professional editor and she really helped with the quality of my publications.

She had a background in biology, so she knew the subject area and she had a lot of experience. I'm not sure how long she'd spent being an editor rather than a biologist, but she had a lot of knowledge of the subject, which I think really helped because she was able to read sentences and actually understand them. So, rather than just from a straight grammar point of view, she could make it actually make sense.

She would go through your document and not only would she make a correction, but she would put a comment in using track changes explaining what the correction was, so if you'd made some grammatical error she would tell you what that error was and why you shouldn't do it. I think she probably had a file with all these common errors and the explanations in them and she would just cut and paste it into your document. So you could learn for next time what it was that you could do better. I found that really helpful. Without that I wouldn't have been able to improve my writing to that degree.

The writing group I mentioned before was started by a friend of mine who was doing her PhD. She was quite keen to set up a writing group because she had been doing reading about how to write your Ph.D. And there was a scientist at a local research institute, and the Ph.D. team was doing some work for her, and we said we why don't we set up a writing group. And the woman at the research institute had, I think, been in a writing group before, so she had some ideas about how to do these things. That's how it started.

We met once a week for a few months—it was quite a long time actually. So each week we would, by the end, generate a "to-do" list for ourselves so we would report back on that and then just share experiences like "oh it was hard for me this week because x, y, z," or "I'm writing a method section this week and what do you guys think about doing it this way or that way?" or stuff like that. So it was quite a free-form thing, just however people felt in any given week would be what we would talk about. Because there was only three other people, it was easy to do that.

We made a commitment to write every day. I mean, it wasn't like we all decided that we all had to do that, but two of us did anyway, and it worked really well. So the idea that they had that I used was free writing first—they were saying, "well you could do it for 15 minutes," but what I found was as soon as I started doing it, I'd only last on other things for about 5 minutes and then eventually it would just morph into what I was working on anyway. So that's why I found it quite productive, because it was like once I'd committed to opening up that Word document on the computer it would just spontaneously turn into doing the writing that I wanted to do. I got two papers written that way.

The group split up because the person who was doing a Ph.D. got to the very end and just didn't have time and was finishing her writing, and then I think the other two went off overseas at different times, and we just never ended up reorganizing the regular meeting times.

Writing does tend to be kind of a lonely thing to do, doesn't it? So just having people to talk to, not even to show the writing—we never showed writing to each other, we were just talking about it—was pretty amazing.

Oftentimes I think it's too late for people to learn to write when they're doing doctoral work. It's hard for people as it is to write a Ph.D. without having to learn how to actually write. And so I explicitly teach writing skills in undergraduate classes, and I do think it's important for people to get some of those skills under their belt before they hit that research thesis level.

I teach writing in my undergraduate classes by making a lot of work for myself which I get criticised for by my colleagues. I use a portfolio, and so for the students I look at it each week (which is where I get all the work from), and I get them to do writing every week, multiple pieces, and all sort of different things that they have to write like more lab report-y things or critiques or answers to questions or whatever. And I always correct their grammar—every time—and I explicitly teach them things like how paragraphs need to be structured. A lot of them don't like it, but usually by the end they say "oh this was really, really useful, and my writing is so much better." I think it makes a difference and I wish I'd had that.

I think if we put people out into jobs, the best thing we can do for them is make it so they can communicate what they've done. I mean you can be the best scientist you want, but if you can't tell other people about it, there's no point in doing it.

I do feel a bit hard-done-by, but I'm trying to make the best of it; it's just finding that way forward. And really, seeing how useful that writing group was, I would like to really do something like that again. We've just got a new chair here and he actually mentioned writing groups. He also said that he expects the senior staff to be reading stuff from other people, and he's trying to introduce a more collegial culture to the department. So I'm quite encouraged by that, and I think that maybe we can do something in the department because I work with lots of really good people, and I get along with lots of them really well, and we have a lot of fun. So I don't think this would be too much to ask of various people, to get together.

To get to the next stage of my career, I need to practice. Get over the procrastination and just practice. I have lots of things to write about, it's just a matter of doing it. I had certain aims about what I wanted to do and I've just found my own way to get there. Which has been slow and I'm not really there. Do you know what I mean? So I don't think that I've been successful—I wouldn't say that I'm successful at all. I wouldn't necessarily say I'm a failure, but I'm a "work in progress"! I'm a *delayed* work in progress.

Chapter 3

TIMOTHY

Timothy is a very senior mathematician, who has moved into interdisciplinary research and is nearing the end of his career. We sit up in his big, bright office, overlooking the city, and he oozes confidence and authority. And yet, as we talk, his narrative changes—of all the people I interviewed, he is the one who most clearly identifies as a struggling writer despite a phenomenal publication record. Perhaps what is most significant about his narrative is the way he portrays his struggles as a writer as emerging from the same attributes which make him an influential and successful mathematician. He is also the interviewee who tried hardest to work through in most detail the process of thinking and writing about his discipline, the way in which equations and language work together.

I DID MATHEMATICS, I THINK, BECAUSE I FOUND ENGLISH SO DIFFICULT

In terms of writing, I was a planner. But that was a long time ago. I'm old enough to remember writing in pen on paper, getting the document together and then giving it to a secretary to type it up. And I often look back at that now with some concern as to whether what I wrote could have been OK, because once I had given it to the secretary there were no changes—or very few changes—made. Whereas now, in the electronic age of typesetting, 20, 30, 40 incremental versions of the paper might take place before we actually get it right for publication.

One of the reasons for so many versions is that scientists do mathematical type setting, and we use LaTeX which gives published quality of text and equations—in fact that's what the publishers use to produce those pages. So we offer publishers the completed books now—basically the publishers have nothing to do to the book or paper other than just add a single line of type style which might change the house style.

Now in that process of producing papers, they're so perfect when they come out that, if there's a full stop out of place, you spend time trying to get it right. You worry about the smallest mistake which in a rougher word processing document you wouldn't be so concerned about. So we're forever correcting; we're forever updating, and scientific publishing tends to go through many corrections, and also often it's on Dropbox as well so other people are looking at it too—they open it up and do some corrections. So it's a very laborious process now by comparison with how I started out. I think in the past we must have done more processing in our heads.

That detail of getting perfection used to be down to the printer or the publisher. That detail of the perfection now is down to us. By and large, in mathematics, we produce the text and we see it as in the final version, as it leaves your machine.

Let me describe my writing process. First of all we might have a theorem that we want to prove. Certainly one would do scribbling and writing it out in some form on paper. Then you would try and commit that—and perhaps the proof of that theorem—to the document. Well, as you know, with word processing you can do little bits whenever you like. That might be the way you do it. You try and get the theorem down and then you build around it. So you'd say, with the introduction, what do I need to say to get to this critical point of the paper? How do I extricate myself at the end by saying how this theorem opens out to other possibilities and its importance?

The other approach, as we're going to do with a paper I'm trying to do with some medics, is to just write down introduction, methods, results. Just write something down in a very brief form which is clearly inadequate, but then you start to expand those areas. Often our work these days is interdisciplinary, so in this particular case with the medics we'll be doing the mathematics and writing up the results; they will be putting all the physiological verbiage around it because I just don't know what it is. I'm calculating something about proteins, but I don't know what these proteins represent and the guy who's working with me, he understands it—he's actually dealing with these things every day. So, you know, we come together in an almost orthogonal way in that paper. Whereas, for other papers, you have some overview of the whole thing, and so you're able to do it all by yourself.

So, oddly enough, you could have papers with your name on where you don't actually know, in a way, what it's about. And you'll often see a paper with many, many authors on. Now that's just making a statement that these people have been involved in some way in getting that result. They may not have even seen the paper. If you have three, four, five people—the chances are that everybody's had some significant role in getting that paper together. But I'm taking the extreme case there where the person wouldn't even know what the paper was, might not even ever find out that a paper was published in their name. You know—it's a little bit like the credits at the end of a TV programme.

And the fact that we can actually move the paper around, as a result of email and Dropbox and all these tools—I've done co-editing on Google with a person on the other side of the world, in Hong Kong. In Google Docs you can both be working at the same time, and you can make great progress that way because people are adding sentences that they feel are needed, then another person looks in and moves it—everything is moving in front of you; some of it is your doing, and some of it is the work of others, but it can work well. This is a revolution in the way we work. When I started, the first 20 years of my professional life, it was a solitary existence. I did the research and papers on my own; I was mainly sole author, but then I discovered that it was quite nice to talk to people!

You know, academics are slightly strange people. I mean, you're putting up with weirdness basically, particularly in mathematics where autism and Asperger's are helpful conditions. Actually you can be better as a result of that, having a little bit of Asperger's—it helps you to concentrate, you know. It eliminates distractions. So a lot of us were loners early in our career—for want of a better word—sad to say that, but I think it's a fair point. And then we've actually found that we quite like working with people, and the facilities have arrived that enable us to have universal word processing, universal access to the files. The grant that we're putting in for the end of this week involves academics from four other institutions. There's no point in us bothering to meet up; we'd waste a full day going somewhere when we can get on Skype and talk about it or work simultaneously. Often somebody takes charge, the paper is with you for the next 24 hours (often it's for short periods of time). We don't say "we'll have it for the next month." The pressures are on to actually develop things quickly and meet deadlines.

It's a general phenomenon that grants are now put together at a multi-institutional level. There's a major push in science for interdisciplinarity, using mathematics in the outside world. So I was just a mathematician who wanted to prove theorems on my own; I still do that in a limited way, but I've got greater respect and money and all the rest of it by going into looking at network structures and working with multidisciplinary teams. And that means we've had to change our ways and become more communal, more extrovert at some level in trying to talk to people and trying to understand their problems, which is not an easy thing to do. You have to spend a lot of time just meeting up, talking about basics.

When we're writing, we have to be more aware of our audience. At one time I didn't even think about it. Again—sorry I'm contrasting my early career with my current career, but that shows up the essence of the problem. But you could just do mathematics and nobody really worried too much what you did. Now they have an interest in you trying to raise research funding and that usually means impact to the outside world, and so we have to think about how we get impact.

One of the problems about working with people who are not mathematicians is that they think that we know it all. We don't—we know hardly anything! You know, I've written books on systems theory. But I could go to a seminar on the subject and not have a clue what the guy is talking about. And that doesn't happen in history, that doesn't happen in geography. You know, I could go to a geography seminar and I might well enjoy it, even if some of the detail is beyond me. I know that I will; I'm interested in geography, though I haven't done it since I was at school. So there is a problem in mathematics that we don't know much and we can't do much, but other people think we know everything because it's a difficult subject. So that's an anxiety that I could be putting stuff down in a paper without full knowledge, which is the usual position of a mathematician, unless they've proved a theorem and they've got the proof of it there; then you can claim full knowledge.

But once you go away from the cosy position of pure mathematics where the theorem is stated, it is proved and it's all over (that is, it's either correct or incorrect and other people can judge it), then you get into the game of interdisciplinary mathematics and then nothing's quite correct as you are approximating the truth in many cases. Do you build a fast track line or don't you? Well, "it depends" is the true answer. Do you reduce the number of proteins measured or don't you? Again, it depends. So, anxiety comes into our profession as a result of trying to work in an interdisciplinary way. I mean, I'm very excited to be working with somebody in a medical faculty. But I'm also worried by it.

Is writing in mathematics persuasive? If I'm thinking about pure mathematics—well, yes, I think it is persuasive because I'm stating it very clearly and identifying the formula. That formula is either right or wrong, and it's persuasive in the sense that people either believe me that I've done it correctly, or they might not have the time to check it. So they have to look at it in a peripheral way sometimes to decide—"do I believe that this guy has got the right idea here and, therefore, as a result of sort of oblique or tangential information, do I think he's actually got the right formula?" Then I think it's pretty persuasive. It requires an act of faith by most people but I still think people should be easily persuaded really as to whether you're on the right track or not; you know, do you have a track record in this area? If you put in simple values, is the answer correct?

Sometimes—it's called an "engineers' induction"—you're only checking a few values and assume it's true for all values. You bring together a combination of things—almost like a doctor looking at a patient: he checks your temperature, he checks a few other things and he decides "well, basically I think you're OK unless you tell me something very specific about a pain here or a pain there." And I think for a mathematician, they'd like to check it precisely, but 40 pages of detailed work? They might not have the time or the inclination and they want to use that result. So, they have to make a decision—"do I use that result or don't I?" If you know that person to have had a decent track record in publishing, you trust their previous publications in the same area, then you'd be prepared to go ahead with it. That's certainly what I would do.

I find short articles easier to write than long articles. As I get older, I don't want to do as much technical checking, so that's a function of age. That extends to mathematics as well. If you give me a big equation now I'm not interested in it in the way that I was when I was younger. I'm interested in ideas and leading things rather than being involved in the nitty-gritty. Again I think that's probably a function of age—and experience—as to what I can do best now, as against what I could do, say, 20 or 30 years ago. I like constructing something as long as it's not too large. For the grant that we got together, I've made a contribution of about three pages, and I had to worry a lot about that, but the whole thing is about 36 pages and I'm glad I wasn't involved with that. I've got younger colleagues who've done that. I wrote books—I wrote five books. One of them took four years to get together. I will never write another book!

I think in terms of developing as an academic writer, up until honours perhaps you don't even know whether you can write or not write—you just scribble stuff down, you don't quite know how it's seen. Effective writing comes sometime after the Ph.D. I went through the book stage which is making a broad statement but requiring a lot of effort which I wouldn't be prepared to do now. That sounds as though I've got a choice. I couldn't, you know, I haven't got the concentration to do that. My mind flits. That's the problem with growing old when you're a mathematician; you've been very sharp at various points of your life and perhaps you understood something that maybe only another 10 people around the world understood; you know you were in that little clique at the very top of that very vast sharp pinnacle going forward. But now, as you get older, you can make this broader contribution. You've learnt a lot. And so you use your knowledge in a different way. So it's research leadership rather than doing the research yourself; helping others to be in the right place to deliver research.

What's interesting is I did mathematics, I think, because I found English so difficult. I did actually fail an important entrance exam; I failed it on English and I was fine on mathematics. I was top in maths but I was desperate in English. I can remember the essay. The title was "My House." Now as a mathematician—and I come back to this autism and Asperger's thing—I've got to write about my house. What is my house? And I went to numbers straight away. It's got five windows, it's got one door—this is age 10 or 11. I knew it was a disaster when I wrote it. But I was incapable of doing anything better.

"The door opened . . ." was another essay I recall having trouble with. I had no imagination in that direction at all; I had imagination in math but no imagination in writing. I didn't realise you could write anything after that—basically that it didn't restrict you at all. I thought it restricted me enormously just as for the other essay with the title "My house." I could have written anything, you know—"my house is haunted" and away we could have gone. I could have written any old bunkum, but I thought I'd got to write the truth, I'd got to write the facts, strive for accuracy. Because accuracy is what mathematics is about.

So I went to a secondary school and my mother, in the first year, wrote my essays for me. Fortunately I had a very good teacher who realised my problem and pulled me on with mathematics in that first year and then I went to a better school at 12. I always knew what was wrong with literacy; I just couldn't always correct it. And when I look back at my early papers, they're not bad. The sentences are OK, they're not too long, they seem to get to the point and then move on; so I obviously got over the crisis at some level, probably by the age of 16.

I wouldn't say I'm a good writer now. I write terrible sentences and I have to correct them. I know how to do that. But, you ask me to write a paragraph and I write it down and then I read it back and I think "oh my God!" It seems OK as it's coming out of my head, but it's not OK when you read it. For that reason I'm worried when I'm talking, that the sentences must be bad as well while I'm speaking because I'm speaking to myself when I write it down. Perhaps I'm not quite as bad as I think, but I do have trouble. It's never gone away; I never find it easy. I show something to my wife sometimes if I'm fed up with trying, and she says "you can't say that." But I do have a revision facility.

I did have a brilliant Ph.D. advisor, a brilliant man, and a brilliant expositor. Fantastic writer. I can remember writing my first paper; he said "is this your first one?" He was a very kind man as well. He said "don't try and put everything in the theorem." You know, I'd saved all the notation to this single statement. So he said "set yourself up for the theorem, put in some definitions of the things that you want to describe in the theorem." I can remember it now, he wrote on pink paper. And he said "now you can write that theorem that looks a monster piece of indigestion in just a single line because you've set up the background." And, you know, that was a critical moment—it took him 10 minutes to tell me that. But I was receptive to help; I wanted help; I knew I needed help. I wasn't a person at that stage who was saying "I don't care what he says—I'm OK." I'm quite sensitive to failing. And so I was ready to take it on board.

I think initially at school you don't need to write mathematics; you do it in forms of formulae; there are very few words you need. So it is an alternative language, and you do feel a release: "I haven't got to do English now, I can do mathematics." Of course I then realised that to actually talk about mathematics, to write about mathematics, you need English and you can't stay within the confines of mathematics.

I think I wrote a decent thesis—even though I'm quite hard on myself. And the early papers look good as well—but that first paper! In my second paper I resorted to the bad old ways of what I'd done in the first paper. And another person had a go at me. He said "oh, you've got to write it better than this." And again he helped me, he was a good writer—and I think that was a further reminder to do better. I knew that I'd got to really look at what I was doing and that other people would judge me as being a bit hopeless, that it wouldn't be accepted for publication—not because of the mathematics, but because it was just poorly written.

I have had these traumas in writing and that's why, of course, I enjoyed writing these books later on. OK, there were co-authors and I tended to supply the mathematics and the first draft, and then the other person was keen to try and understand what I'd written down and we reconfigured some of it. He probably wrote the majority—I would say that, out of five chapters, I did about two chapters' worth on my own and he did about three. I thought it was good to have somebody around to check.

So I think it was that first year in my secondary school and the quality and the help of the great advisor that got me through basically. It's been a big thing in my life.

But I'm pleased that I've got a job where I have to write. I guess that I could easily have taken a different path—I might have enjoyed being a plumber. But then I would have missed out completely on writing if I hadn't been lucky with my teachers.

What I continually emphasise to the students here is that they too possibly think that mathematics is an alternative language to English, and a lot of them choose mathematics, particularly from abroad, because—"oh I'm not going to do very well in English" or "that's going to be my second language" or whatever (although some people with English as a second language seem to do better than the indigenous population, but let's put that to one side). They say "I can use mathematics—I can get by with mathematics." And my entire emphasis is you must write; you must write about it in English. So you must put words around the equations. Just putting the equations down means nothing.

I said it this morning in my class at nine o'clock. "Write something down to show that you understand it." And they don't like hearing this. I said that's the skill; never mind whether you've done hydro-dynamics or calculus or algebra—it doesn't matter. How do you explain it? How do you write it down? How do you communicate it to other people? And I think that's what they've got to get. If they don't get that, they miss the point of doing mathematics. But they just don't get it. The penny has to drop in all sorts of ways for all of us, and it can take many years. The simplest things that we have to learn in life often take a long, long time to appreciate. You understand what the person said, you get what they say, but you just somehow don't believe it until it happens. So, we don't believe even simple things, do we, until we've actually experienced them?

I would say writing is not separate from mathematics but part of mathematics in the sense that I also say to these students, "when you're looking at this page, can you talk to yourself about this? You need to be able to tell yourself what you mean—you need the words." If I write $F:R \rightarrow Z$ and I can't explain what it says, then it's not much use. If I say "we have a function F from the real numbers R to the integers Z," the words are actually reassuring me that I know exactly what I'm talking about.

The equation is saying all sorts of things: 1. I recognise that this is a statement about functions; 2. I recognise a function maps from one set to another set; 3. The set it's taking values from is the real line in a set of real numbers and by applying the function f we obtain an integer. Now that's telling you everything about the setup by using those words. Whenever you can use a sentence you should use it, even if you don't write it down. So I would say that, coming back to your point about creativity, when I'm creating the mathematics, I am talking to myself, I'm saying "I'm looking at this, so I've got to look at the graph; I've got to find the maximum of this graph; perhaps I've got to find out where the slope is equal to 1." Now if you can't say those simple phrases it means you can't think about the problem in front of you.

So you have to talk to yourself about the mathematics while you're creating it, but then when you write that's a separate process. So when I'm doing a scribble about a calculation here, and talking myself through it, I'm not writing at that point. But then, perhaps what I write would be very different from what I said to myself; I don't need to write as much as I've said. The English has to be used the entire time, but it's used in different ways while you're working with the mathematics and while you're writing it in the paper. Does that make sense to you?

Let's see if I can explain it better. When you're working, when you're conceptualizing, you've got to be able to put words around it; but perhaps they are short phrases to compose a thought or a question. But then, once you get to the writing of the paper, then the phrases have to be composed to make meaningful sentences. So when it comes to writing the paper, I might be writing completely complementary things from the sorts of processes that were going through my head while I was creating the mathematics itself. It's not unusual, when you're writing, to discover a new idea or something you hadn't thought of originally; I mean, you might find a mistake in what you've written down. Perhaps you realise you need to put an extra step in; while you're writing it, you might actually see that it's not true anymore.

I do like to read. I'd like to be able to read novels freely, but I don't. It's easy for me to walk away from a book. I see people get into the first five pages and they can't stop reading. There have been occasions in my life where certain groups of books satisfied me like that, but only a few; as a child I couldn't put *The Famous Five* books down. But for some reason there was nothing else that captured my interest. I like biographies, so again I'm tempted towards factual things. I'm interested in astronomy, I'm interested in all these things—it's not that I don't have an interest. I find it difficult to concentrate—I can read on holidays and I can read when I'm officially switching off. The thing I read most is the news. I read a lot of newspapers. When I read, I want it to be based on fact, I want to believe in something, so I have also have difficulty with sci-fi if it looks crazy.

But the other problem is mathematics is my hobby as well as my job, and I probably do about 55 hours a week. So I've got enough to do basically, but that's not an excuse. That's just how it is. But—final example—I've been to see Shakespeare quite often, and never really understood it—my wife has to tell me what the story is or who's the mother of who or whatever. I've been to *The Tempest* three times now; I know it's on a desert island, I know there's a storm, but if you asked me one other question about it—I wouldn't know the answer—that's disgraceful, but true! I just switch off. I just sit there and just watch it. But the Shakespeare I did at school—*Merchant of Venice, Macbeth*—great. I liked those stories because we kept on going through them over and over again and eventually I could get some of the nuances in the plot. I have to be in the right state of mind to follow the plot. My mind seems not to be constructed in the right way to follow them.

MASON

Mason is a solid state physicist and seriously annoyed by me. During our interview, he objects to most of the questions I ask, suggesting they are either irrelevant, ignorant, or biased towards a humanities' perspective. I struggle for a while with my own impatience, until I decide the best approach is to simply let him explain what *he* thinks is important about science and writing, without being tied to my questions. And he surprises me. He was the only person I interviewed who expressed these views in such uncompromising terms, although echoes of his perspective can be found in some of the emerging and doctoral scientists' perspectives.

WORDS AREN'T IMPORTANT

I start writing at the beginning. I start with the title, list of authors, then I may get stuck in straight away to the introduction; but the writing isn't actually a very large part; preparing the figures is the major part. Sometimes I do some of the figures early on. But really I start at the beginning and go to the end. I've usually got a pretty clear idea of what's going to be in the paper. Because you do the research, you get the results, you say "this is publishable"—so you write up. It's quite straightforward. It comes out fully formed. I don't have to do a lot of revision if I'm writing it by myself. If there are co-authors involved, obviously they then would do some editing, make some changes—and then I might want to make more changes in response to what they've done. But if I'm a sole author, then I get the final form—until it's been to referees, that is. The final form isn't very different from the first draft.

If I'm revising stylistic issues, I work intuitively. If it's actually a matter of scientific content then obviously it's deliberative. Saying "this argument doesn't hang together" is scientific logic and needs thinking about. If it's stylistic it is just a question of how it reads.

What you call "structure" I'm not giving a name to. It's impossible to make a distinction between what you call "structure" and style. Style appears at all levels. Even though someone would have written a paper differently, their personal style is different from mine and that shows at all levels. The content is another matter. You see, scientific papers actually have content, and you might guess that I suspect that's not always true in the arts and humanities.

So, you take a math proof; it is an entirely different matter whether each step leads rigorously to the next—that's structure. Style covers such things as how you choose to lay it out. If there are parallel paths in it, how you prefer to set that out may be different—two different mathematicians can present the same proof very differently because they have different styles. But it's the same proof. The content is the same. Everything which is not content is style, which belongs to the person. All the content is objective and comes from, if you like, the experiment or the theory.

I'm not quite sure if I think about the audience, but I certainly think about what I would be saying—because, after all, a great many times a paper is connected with a conference presentation and for a conference presentation I would certainly be thinking "what am I going to say?" Now that, I suppose, implies an audience because I would never read it aloud in a room without an audience. I don't do popular writing and even with what we might call interdisciplinary stuff, or specialised stuff, I believe in making it as clear as possible to any colleague. I haven't tried writing for undergraduate students; I write for people who are interested in and knowledgeable about the subject. And if they're not knowledgeable then there are references for them to find out what background they need.

The physicist and mathematician spend all their time adapting to new results, new information, new problems. Whether they're writing in new and different ways is not something I've really thought about.

I think *convincing* is a word I'd rather have attributed to my writing than *persuasive*; persuasive carries a slight connotation of trying to sway people by emotion rather than by the logic. Maybe convincing does too a bit. I'd prefer the material to be convincing rather than the way I write. The figures, of course, are crucially important in any scientific paper in terms of conveying what it is you want to convey, in terms of actually telling the audience something. In that connection, I must say I learnt a lot from a French professor, a very great man, very eminent. I didn't speak French, but I went to a research seminar that he gave in Paris, and I was able to follow completely what he was saying. I mentioned this to him afterwards and he just said "that's a tribute to my diagrams." He said explicitly that for writing papers, what he always does is lay the figures out on a table, make sure they tell the full story, and then throw some unimportant words around them.

Words aren't important in science or engineering. I've had this discussion with philosophers frequently and they claim they can't think without words. Of course every scientist and mathematician and engineer thinks without words. And if you ask why use the words, it's like using prepositions in writing—why do you use prepositions and things, linking words that intrinsically have no meaning? The words, if you like, are the prepositions between the figures.

I would never read a paper from beginning to end, absolutely not. I would ask "now, what's he actually saying? Oh yeah, now why's he saying that?" and I don't need to read the rest. I go to the conclusion and then to the graphs. The conclusion was what I thought was important; and I might find the conclusion is sufficiently well stated in the title or at the end of the abstract.

You might say a good writer is one whose work goes sailing through the referees every time, whose papers always are published and accepted as first submitted. I'm not so sure that's true because in my experience it's the boring papers that have very little new in them that sail through. When you're sending in anything new, the referees are a random selection from the potential readers and every referee would make different objections and raise different difficulties. I wouldn't say even your person who rated 10 on a scale of writing competence would be able to preempt all the referees. So I don't think the measure of a good writer is whether a person's papers go sailing in, I think the measure comes later in whether people find it easy to grasp what quality the paper is; I don't think my readers have any difficulty doing that with my work.

I find it easiest to write a paper with a very clear central idea and good data to support it. But I don't think there's very much difference in difficulty between different kinds of writing—a report to a client is simply a scientific paper that's not going to be published, and a grant proposal is (until they started putting all these extra requirements in sections on impact and so on) very much the same except they're in the future tense instead of the past.

I write because there's no point doing research if you don't publish; and research is fun. But you can't just do it for fun and not publish. Who'd pay for it? No, there is a famous physicist who did that—he's the one the Cavendish Laboratory in Cambridge is named after—Henry Cavendish.[6] He was about 20 years ahead of his time. He got most of the results that other people got later but he did this purely for fun and he didn't publish anything. It wasn't known until after his death that he'd been doing it! Well that's crazy. He was a rich man and did it as a hobby. I would have thought there would be a totally natural urge to tell other people about interesting things he'd done if only going down the pub and saying "you won't believe what I found."

Scientific writing evolved out of correspondence, out of people like Galileo and Newton writing to other people who they knew were interested in telling them what they'd done. So growing out of that I would say the writing isn't part of doing the science; it's part of telling other people about it. But, that's where figures and math come in because, of course, they're both involved in putting pen to paper and they are both part of doing the science. So doing the graphs is part of the science, doing the figures, and seeing if the theory lines go through the data points—and being very pleased when they do. That's all part of doing the science.

Does writing help develop new ideas? Your writing could do that I suppose, or you could say it's the figures that did it. Or it was the referee who didn't understand something and needed another figure. Where you get ideas from is a completely different question. I don't particularly get ideas from writing. You get more ideas through sitting in a pub with a pint of beer, staring at the beer mat.

I have no urge to start writing for the government; I've no urge to write books. I've only written a couple of review articles, and both times it was because someone twisted my arm. As for books—I certainly learnt when I was a Ph.D. student that most of the famous textbooks are written by people who haven't done any research for quite a long time. When you can't do anything original you start writing review articles.

The other point was about this phrase "learning to write." Some people might do that, like some people actually have to learn to spell. Other people don't need to learn to write because they do enough reading.

Solid state physics overlaps with a lot of other disciplines, and we actually find it relatively hard to communicate with those other disciplines because their style is so different. One of my colleagues found it very difficult to write a paper with a chemist as a co-author because the chemist wanted to put in all sorts of things that the physicist thought detracted from the paper, detracted from the point. So, it's hard to say what is distinctive about the style of a solid state physicist rather than a material scientist, but it clearly *is* different.

Too much detailed knowledge is something a physicist never has, because physicists are fundamentally lazy. I tell the students that we need to have the right amount of laziness. Not so much laziness that you don't solve the problem, but enough laziness that you actually put a lot of effort into finding the easiest way to do it.

I didn't like writing as a child. I don't think I was taught the basics at school either. I only took writing-based courses to the age of 14. I dropped them as soon as possible. We did very little writing in science at school. I learnt to write science by reading, reading broadly. As for how to write a scientific paper, as soon as I started reading I was immediately learning how to write and that would have been when I started my Ph.D. I'd say my Ph.D. advisor didn't teach me how to write—not in the slightest. I mean obviously he wrote some papers with me as a co-author, but I didn't see any particularly different about the papers he was writing than the rest of the literature I was reading. It was almost effortless to learn to do this.

I don't think I have a role of teaching writing. With my Ph.D.s, I'll push them out of the way and get on with the writing myself if necessary, because I'm not going to sit over a text with them explaining to them why you have the word "introduction" in bold here and what it is that follows it. But when the question does arise I always tell people to Google Boxman Tel Aviv.[7] I was at a conference a few years ago where this guy was giving a lunchtime talk with free sandwiches and, out of curiosity to know what he might be talking about, I went and had a sandwich and what he was giving was a lecture on how to write up a paper.

So if you want to know how to write physics go and read Boxman—you don't need me to tell you how to write a good paper, Boxman does that. And the point he makes is that each section and paragraph should actually have that same structure. And I've often said that really out of the 80-odd slides, slide 18 is the only important one. I often lay it on its side and call it the jet engine model of writing, because if you lay it on its side then you've got the air intake on the left, you've got the reactor where the work's done in the middle, and then you've got the cone at the back which actually makes the output do something useful.

Oh, and as well I think I'll say to them—for example, yesterday I had a new Chinese research student preparing a talk for a conference and he wants to know whether his outline is suitable. So I glanced at his outline and I said "what do you want the audience to remember the next day?" And he hadn't thought about that. If I have a Ph.D. student who comes back with a mess, I point him to books.

I don't actually enjoy the mechanics of writing, it's the product. So the question of liking to do a kind of writing, the actual process of writing—whether pen on paper or with a keyboard—is in itself not very pleasant. So I only want to write things that need writing.

Words have a purpose but that purpose is less clear than the purpose of scientific symbols and mathematical symbols in writing. I guess the question you haven't really asked is what language do I write in. Accidentally English, but, of course, that's just because the entire scientific literature is in English these days, but it wouldn't be any different if it was another language. But we've learnt to write in it, perfectly competently, just by reading it—this is how the Chinese can write papers in English—not that they speak English very well, not that they're fluent in it. But it's a language which is scientific English, which they can learn. It's no harder than learning algebra. Writing academic English is much easier than writing conversational English. I mean, if I wrote conversational French, then the French would laugh at me; if I write scientific French it's a bit of an improvement, and if I actually wrote very much scientific French it would soon become indistinguishable from a French scientist's.

The language is simpler. It needs to be, just because it is international. While there's no way I could write like a French novelist—I could read any amount of that and not be able to write in that style because there's far too much that I didn't learn through not having French as my mother tongue. But scientific French is absolutely no problem. Same with reading. I examined a Ph.D. in Spanish and I don't speak Spanish. I arrived there—I'd got the thesis with me—and I spent the entire afternoon on the beach reading the thesis in Spanish. And it started off rather difficult, and I was relying mainly on the figures, but after three or four hours, it's so repetitive, the same use of the same structures, I was actually reading it quite fluently. Then I went to the defence the next day, and the student had told me that he would start with a minute in Spanish and then switch to English for the rest of the presentation, and after his minute in Spanish I just could not pick out a single word of that; I couldn't even detect the title of the thesis—and this after spending the whole of the previous afternoon reading. That's the difference between the scientific writing and the rest of language.

CHAPTER 4
THE WRITING COMMUNITY

> I think of historians as lonely people who aren't necessarily surrounded by as many people and the opportunities that I have to work out their ideas as they're going on.
> — Gao, Chapter 4

> Science is, of all the creative areas, the most social. You take a writer—they're a very solitary person. A composer, an artist is solitary too. But as a scientist, it is very hard to be solitary. . . . It is very social.
> — Richard, Chapter 2

One of the most significant issues that emerged through these interviews was that, although writing in the sciences is largely a collaborative process, it can be, at the same time, an isolated and often *isolating* experience. From the graduate students who often struggle alone with the writing process, to the senior scientists who make a switch to cross-disciplinary or public-focused writing mid-career, writing can be a lonely activity which is rarely explicitly discussed with colleagues.

Some graduate students are lucky. Those who fare best tend to belong to a large research group, working on a project that is funded as part of a larger project, led by a senior scientist who has the language and the motivation to talk about writing. But most of the doctoral scientists I interviewed—and this was confirmed by most of the interviews from senior and emerging scientists speaking of their own experience of developing as writers—perceived writing as not explicitly discussed with seniors or colleagues. And this lack of discussion was ongoing, into the collaborative work of senior scientists:

> . . . we didn't talk about writing, we talked about the math. Not the writing process, ever that I can think of. I mean we didn't have typesetting to even waste time talking about that which is now what we talk about when we're not talking about math. (Senior Scientist, Interdisciplinary Mathematics)

For some emerging scientists, as we have seen in Chapter 3, this lack of intentional, articulated support is traumatic. Those who struggled most were the doctoral students whose project was self-standing; it seemed that, in this situation, advisors and peers showed even less capacity or motivation to discuss writing.

This lack of discussion was puzzling on a number of levels. Most participants in this study saw writing as, in some way, critical to the work of science. Everyone agreed that writing was a way of communicating findings, but for many it was far more than that: it was *part of the work of science*. It was the way a scientist tested their ideas, interpretations or intuitions. It was the way new ideas—either about the current research or new research—emerged.

The literature on effective research communities seems also to suggest that issues critical to the aims of science, such as writing, should be discussed within that community. Bereiter and Scardamalia (1993), for example (see Chapter 1), consider scientific research communities as the archetypal learning community. The notion that these research communities are not engaging with writing *in the same way* as they engage with their science is therefore intriguing.

The answers given by participants when this issue was questioned became reasonably predictable: while agreeing that writing at advanced levels should be taught within disciplinary communities, most scientists felt they were ill equipped to teach these skills. They used these skills themselves, but because they had learnt by reading and imitation, by gleaning meaning from their advisors' or co-authors' revisions, rather than through conversation, they felt they did not have a language with which to discuss writing.

Another way of addressing this question is to hypothesise that, since science and writing are not separate processes, scientists are, in discussing the science, implicitly discussing the writing (Graves, 2005; Yore et al., 2002, 2003). The second narrative in this chapter supports and demonstrates this idea: in discussing his ideas in multiple venues with a range of audiences, Gao is experimenting, not with the data itself, but with how best to construct the story that will convince an audience. His process calls into question Yore et al.'s (2002, p.689) finding:

> [Scientists] appeared to disconnect the verbal interactions about their writing and the embedded ideas as being part of the writing process. . . . It is hard to believe that a team meeting to consider draft reports or to address reviewers' comments about style and content did not focus their discussions and clarify their thinking about their understanding.

However, the point remains that most scientists in this study perceived conversations about, and collegial support for, writing to be in short supply—*and this was felt as a lack*. A worrying number of scientists, both emerging and senior, did not feel explicitly or sufficiently supported, engaged, or trained as writers of science within their research communities.

In the course of my research, I encountered one group that I felt approached writing as a learning community. The two leaders worked within the same institute and were friends, and the group consisted of faculty and graduate students. There were two catalysts for this group. First, the leaders were already concerned for their own productivity and for the writing of their graduates. Second, these instigators attended a writing seminar where they acquired a language with which to talk about writing. Already motivated to write, but struggling to find a way to write within their busy schedules, this seminar gave them practical strategies which they found convincing. It is significant—and an important point for writing teachers working in the sciences—that their professional scepticism meant that they needed evidence that this would work, and they also needed to test empirically their own experience and the experience of others within the writing group.

This chapter opens with two narratives which highlight how an individual is supported in their writing and their science by engagement with others. Mostly they write alone, but their community is critical to that process. The first story is that of a doctoral scientist, Eugene, who focuses on the diverse nature of the community with, and in which, he works. Gao, like Eugene, is active in seeking out support—though his support network is perhaps more traditional in the way it comprises peers and academic mentors. The chapter concludes with a narrative from one of the instigators of the writing group and a conversation between the two of them. For these scientists, developing a writing community has been a solution to particular problems but also a cause of concern: how can they maintain the momentum?

EUGENE

Eugene has a gift. Within a few moments of listening to him, I'm so interested in his extraordinary migrating birds that I struggle to stay focused on writing. I've since watched him weave the same magic in public seminars and listened to him convert a sceptical radio interviewer into an instant conservationist. What he conveys is that his research aims to explain a genuine mystery. He has travelled around the world to research his birds, and for five years now, he has bunkered down on the beach that 'his' birds fly from to watch them leave on their long trip. He appears to know each of these hundreds of birds personally. His views on writing are perhaps unusually sophisticated for a doctoral student, and are informed by his first career, as a graphic artist, as well as some superb training in scientific writing experienced earlier in his career. But the most distinctive aspects of his writing are its relational quality, and the diversity of the community that supports his work.

Chapter 4

I Don't Just Have My Advisor—I Have All these People

When I was going through my master's degree, I noticed that I could split the master's researchers into two groups: people who were really serious about research and people who weren't; people who were getting a master's degree just to get one and people who were really interested in doing research. And the people in the category who weren't that serious tended to be people who were afraid of writing and not confident about writing and didn't want to do it. The people who were really into research were much more likely to be good at writing and confident about writing and just be able to write. It is just an observation—I don't know what that means exactly, whether writers are predisposed in some way.

I find people to be atrocious writers in general. Most people are not into it or are afraid of it or something. Even in classes during my master's degree, when we were doing scientific writing, I was amazed at the low level of writing, the lack of ability to put coherent sentences together. Maybe it's more about thinking than it is writing as far as putting coherent thoughts together. And yet, in scientific writing, it's so logical, it's just like A B C, compared with creative writing.

Scientific writing is very cold and impersonal, and I do a lot of it. But the writing I enjoy a lot more is much more informal. During my research in March, I'm super-busy; I'm doing field work every single day, and I get extremely isolated from people. I could go the entire month without talking to anybody because I'm in the field almost all day, come back, collate the data and write notes. So I found a way to handle this almost in a social way. The project I am working on is on migratory birds, and there's a really hard-core group of people spread all over the world who are interested in this topic very intensely, and we collaborate and communicate. It's a very tight group of people, a lot of whom have never met each other, and my results are pretty interesting. I thought one way to communicate with people and get my thoughts together and have other people respond to what I am thinking would be to publish something like a blog of what I am doing in my fieldwork. So what I do during March—I've done it the last two years—is basically rattle off a three-or four-page report of what I have done that day and what I have found and what it means in my study, and I send it to people all over the world. It's a list of like 20–25 people who are really intensely interested in the subject and I treat it very informally. Sometimes it diverges from science quite a bit. I mean the science is in there and that's the basis for it, but I seek to entertain myself and other people by just going off on surreal tangents and saying things that are blatantly false, just to stimulate discussion.

Since I've become a researcher, this has been the only creative writing outlet that I've had. But I've enjoyed it because it's a sharp break from the scientific writing where you don't get to interject your personality much at all. I needed something that was my personality coming out. So it was a lot of fun and people seem to enjoy it. It's like keeping a research journal but with an audience.

This social aspect also turns out to be important for the scientific writing, too. Let me give you an example from today as I was trying to write—I haven't written a single paragraph today, by the way. There's a lot of research being done on migratory birds by non-professionals, volunteers, and a lot of this stuff is sitting around in very capable hands and not being published. And so today there were some questions I had about the Australian population of the bird I study, and there's a lot of data on comparing the Australian and New Zealand birds, and I know there are people out there that know the answer to the question that I'm wondering about. So I'd simply send an email out to five people and say "here's my question—I'm doing this with the data and these are the things I'm finding interesting. What's your information on that?" It usually ends up with a two-week email conversation, and people start adding people from Europe because they say—"oh, that guy knows something about that—here's his email." And so it becomes this ever-widening group of people discussing the question I have for one paragraph in my dissertation. So it's actually really interesting because I don't just have my advisor—I have all these people.

Bird-watching is huge and there's no clear division between bird-watchers and scientists. A lot of the best science is being done by people who are doing it in their spare time because they love it. I don't think there's much financial backing of some of these things here, and so it falls into the hands of people who just have a love for the subject. They're retired people and people who have jobs and are just doing it in their spare time. I'm absolutely amazed at the number of people who are in this group—like, "I'm a house builder, 60 years old, and just decided to start cannon-netting birds and recording data for 25 years in my spare time!" I mean, I've been researching this bird for four years but this guy's been researching the bird for 25 years and he's never been paid, he's got no degree and he's not going to publish his stuff because he's not a writer, he's not a scientist, but he's got all this stuff and he's this incredible resource for academics. And my advisor, he's also part of this whole thing, and both of us benefit continuously by contacting these people and building relationships with them. And they're all great people because they love what they do and they're really excited to share it. Academics in a similar position might be more guarded about their results and their thoughts, but these guys don't even care about publishing or being a co-author—they just want their work to become common knowledge. It's a great position to be in; I have more than 20 years of experience at my fingertips even though I've only been on the scene for four years.

Chapter 4

I finished all my fieldwork in April when the last bird left on migration, so since then I've just been writing and analysing data. I did three field seasons and during the last one I was actually already starting to write, which is a little hectic, but I'd gotten two manuscripts submitted already before my field season ended. So my Ph.D. is going to be basically six published papers with a general intro and a discussion added onto them.

Each one of these chapters has to be a completely coherent paper that a journal is interested in, so there's a bit of repetition going on there, but still I think that's a better way to go than to write a thesis and think that you're going to go back and rewrite it for journals, and then never do it, which you see time and time again. I've finished writing a short methods paper which I thought would be very interesting to people in the field, so I got that out really quickly and that's already in print. Then I had a second paper which we also rushed out early because we thought we had a very sexy, interesting result that would get in a very high-level journal. We wanted to get that out quickly, not only so other people didn't scoop us, but also because if we got it in a high-profile journal we would be able to use it as ammunition to get post-doc funding. So while I was doing field work, I got this thing together, a short paper that we submitted to *Nature*. It was rejected without being reviewed, and then by *Science,* and then two weeks later *Nature* invited us to submit it to their online journal—it's called *Nature Communications*—and it's been accepted and I just put in the last changes.

And so I've gone through multiple rounds of revisions. During that time between March and now, I've been working on the third paper and collating and analysing data for the other papers as well. So I will have six papers and I'm in the final throes of this third one. In the next 3 or 4 months I need to write three or four papers. A lot of the data is sitting there waiting to be written up—I have tons and tons of data and to figure out which goes into which paper is a tortuous process. But even though this one paper's taken me way too long, the path seems very clear to the other ones. So they should come out very quickly. It's not like a retooling for every single paper. They're very interconnected—sometimes I'll write a manuscript and say "oh, that part belongs in that other paper" or the editor will tell you that. But right now, I've got all these pieces and it's a matter of putting them in the right spots. So that should get much quicker.

The first thing my advisor and I had to do, before I even came here, was collaborate to write the proposal for the Ph.D. grant, and we did this by email without ever talking to each other; that just went back and forth and we sort of wrote it half and half. So when I got here I had an advantage over many Ph.D. students, in that I already knew exactly what I was going to do, rather than spending months developing my project. And in fact I started collecting data as soon as I could buy a car (to get out into the field). We basically knew what needed to happen and so some of what I did was backwards compared to other Ph.D. students, because my official Ph.D. research plan was presented to the university after I already had a field season of data. Which in fact is the same thing I did with my master's degree.

I would say, in my experience, that the idea that there's an *a priori* hypothesis, that you collect the data specifically for it, and then report your results, almost never happens in wildlife ecology. People don't know what they're going to find, and they don't know how good they're going to be at finding it, and they're scrambling for sample sizes no matter what. So usually it's a matter of "that's really interesting—let's go find everything we can about it" and then later on you think "well, I *could* have had that hypothesis—it certainly sounds believable. But that was one of the ten hypotheses I had and it's the only one that works so we'll write about it!" It's not strictly hypothesis-testing but sometimes it's dressed up as that.

In terms of working with my advisors and getting feedback, my second advisor was brought in as a mentor of the whole process and it's turned out very good that way. I mostly work independently and then just give drafts to them and they're both really good about turning things around. A lot of times I have given them pieces that are just fragments of stuff—like, "here's a write-up of the results" and then just a schematic of the intro and the discussion including what points I might hit. We meet very regularly—every week or two—and go through ideas. So, the thought process has always been peer-reviewed and I never go off for three months and then come back and say "look at this" and they go "*what?*" We are really collaborating on what the thought process is at every moment. It's a good process. Both my advisors are really energetic people and they are both extremely interested in my project. It's been a great collaborative experience.

The relationship has not been such that I give them a draft and they hand back the entire draft with all their comments written up on it. We have been working in the form of having a meeting where they print out a document I've sent them and scribble stuff on it but they usually don't just hand that over for me to interpret on my own. Usually we go through the manuscript together and they say "this is unclear" and I'll go "OK—I'll make a note of it and fix it." Because we've already taken a couple of things all the way to publication, there have been times when they have taken a word file and tracked changes and actually rewritten sentences saying "I think this would be better," but that's not usually the case.

I had lead-authored four papers before I got here and so I think, to a certain extent, my process is more refined than a lot of Ph.D. students. I don't think I'm being taught anything specifically about writing—I mean, of course I'm getting better at it through practice and because my project is much more complicated than when I was doing my master's degree. But I'm not sure that I've learned anything specific during my Ph.D. that I can put my finger on about "this is how you approach writing." There are stylistic things that I used to do that I don't think are the right idea to do anymore. For instance, there are different ways to write papers and one of them is to write it linearly, but I'm finding a better way to work is to write it backwards.

Actually, I was having this discussion a couple of weeks ago, with someone who's just starting to write. What I think young writers do is they write their intro first and it becomes this big blathering of everything they've learned on the topic and that's not what a journal wants to hear and it's not what a reader wants to hear. And someone taught me this way back, that you should write your results first, and then write the methods that explain your results. You don't write everything you did—you write only the methods that explain what you want to talk about. Then you write your discussion, which doesn't talk about anything except the results that you presented, and then you go back and write the intro that would lead you into that. That's very different from writing the intro by talking about the subject, especially when you're given the task of "do a literature review on everything that has to do with your topic." You have all that at your fingertips; you have all this literature that you've been thinking about and so you start to write an introduction and you could just go on forever because your general topic is huge. No matter what specific thing you're studying, the topic is bigger than that, and so you end up with big, bloated rambling introductions that actually don't set up your paper very well.

Right now I'm writing a very complicated paper that has data taken from all over the place and different sources, and it's been a nightmare. The methods are long because it's a bunch of different types of data, and the results are longer than you'd expect. I wrote a three-paragraph intro for it and I'm totally happy with it. Because I realised that that's all it needed; all they need to be told is why I'm doing it and why it's relevant—anything else can be left for the discussion if you still think it's important. I think that's the biggest transition I've made, almost out of necessity to be more efficient, to write backwards.

What happened in my master's degree and my first paper—and I think I see this a lot with people writing their first paper—is people want to say *everything* because they're proud of what they've done. They've done a lot of things and it's all cool and they want to get it all in there. And with my very first paper I did that. I ended up writing three more papers on a related subject and by the time I got to the fourth one I looked back at that first one and thought "what was I doing?" But that first one got published in a nice journal anyway.

Drafting that paper for *Nature*, there was something about it that I really loved and when I finished the thing I thought "that felt great." And the reason it felt great is because it didn't get into all the grey areas that bog down ecologists. The word limit was ridiculous—it had to be really short; we knew that it couldn't ramble on. So I would write one or two sentences on a subject that I knew I could write an entire paper on. You know, I could write 4,000 words on that topic and I even had data that would confirm what I said, but I knew that *Nature* wouldn't put up with it, so I was just writing "dah dah dah—this could mean this—*end*," and then moving on. It was really liberating to write a paper that way because ecology is so complicated that if you're really thinking about what you're doing you'll never be able to write a sentence. It's so muddled with "maybe" and "these guys found something else" and "I don't know what that means" and "it could mean these twelve things"—but when you're writing a paper that short, it was so nice to just put that all aside and just say "what do you know?" and just write it and stop. It was criminally short, but it felt so liberating because there was no way I could get bogged down in the things that normally bog me down.

But while it was liberating, it was also unsatisfying because I kept having to put aside things that I knew or just say "this might mean this" when I knew there was a lot more to say about that, which was also kind of painful—it actually felt a little dishonest. But when the reviewers' comments came back with "what about this?," "what about that?," I was right there with "well I have all the data on that—it just wasn't in the paper." And so I found the best way to respond to the reviewers, because I didn't have the word limit any more (*Nature Communications* doesn't have the strict word limit that *Nature* does), was to actually make the paper about 35 or 40% longer. And at the end of that process I think it became a much better paper—at least for the purposes of my dissertation because now it stands better as a chapter than it did in that ridiculously abbreviated form. It feels like a more honest paper—I feel like I'm not hiding as much stuff; I had the length to actually deal with some stuff.

It's unusual for a master's student to have four papers published. This is what happened: I attended one of the best wildlife programmes in the US, and one thing that they really did was make scientific writing a focal point of basic wildlife classes that were being given to bachelor students. So, my very first course, my first semester, included an in-class research project and they had a policy that across all their courses any of these research projects were written up in journal format in the style of a given journal. This was a brilliant move. I think this is probably the single best thing that has happened to me as a writer. So in the first class (this is also being taken by second-semester bachelor students), we were given a file which contained the publishing guidelines for a particular journal. And so we had to write abstracts and everything; we were writing in essentially the same format I am using today. I used it right from the beginning, so in two years of course work I'd written probably six or seven reports in journal format.

The faculty were always publishing stuff with their grad students, and so you were getting exposed to real scientific writing and peer-review in your first semester; it was absolutely brilliant! Many of the issues around writing things in a scientific manner were just dealt with right off the bat and were already second nature by the time I was writing up my master's dissertation. And I can't thank them enough. We would do class peer-reviews where we would break off in teams of three or four, go out and do field research and we would be writing collaboratively with two or three other students on a report that was in journal format, and then we would bring it into class. Everyone would peer-review two other people's manuscripts. All this stuff about the economy of scientific writing—it was painful. The only previous writing I had done was creative writing, and the extremely terse and to-the-point writing that is required in ecology was beaten into us immediately. We would turn in these bloated manuscripts and the teacher would say: "All this is unnecessary. These five sentences could be said in five words." And just right from the get-go you were starting to think how to be super clear. And that is so much more important than anything I've learned here.

I think that's the way scientific writing should be taught. We didn't have any classes that were about scientific writing. We didn't have any classes that were about writing at all. But every single class had writing in it.

Do I think about my audience when I'm writing? That's a good question. Maybe not as much as I should, because I personally don't have really strong relationships with particular journals, and so I often don't even know who is the typical person that's going to read that journal. I guess to a certain extent I don't know who my audience is in a lot of these cases. In some cases I do, like the methods paper I wrote where I knew specifically who was going to read that and what they were going to be interested in. So that was very specifically written for specific people, and it went to a journal that I knew would reach them.

I think anyone writing their first scientific paper is probably uncomfortable with their audience, and so when I started out writing papers I didn't know exactly how what I was writing was going to be viewed. "Am I a total poser? Am I just going to be outed as someone who doesn't know what they're talking about with what I've just written?"

Science is supposed to be all unbiased and matter-of-fact, but we know that's not exactly true. So in a sense, I guess I am trying to be persuasive because of course *I'm* convinced of things; hopefully I'm not convinced of them before I take the data, but that's not always the case. But after I've taken the data, I'm probably convinced of something and then the whole process of writing gets mired in the statistics which I'm very cynical about. I mean in some cases I don't know, and so I will write a few paragraphs basically saying "I don't know and neither do you—you know, these are the messy facts and this is just the way it is." And other times I'm pretty convinced of something and then find the best way to make sure that comes across. It's always a tricky line to draw between whether you're just trumpeting your pet theories or whether you actually have facts behind you, and so that's something that has to be constantly thought about and hopefully your peers call you on it. I try to be as objective as possible but I don't think it always happens.

I'm not sure whether I just have a contradictory personality or whether this is just the nature of scientific writing, but I've found that my papers tend to say "this is what people have believed, and they don't have the evidence to believe it. Here's some evidence; what I've shown contradicts what everyone thinks." And maybe that's the only way you're going to get published, because if you just write "I took this data and it agrees with 100 papers before me," no one's going to care. But I wonder if I actually exaggerate the combative nature of it just to make it more interesting to read. Because I do think that, to a certain extent, scientists (and this is my big pet peeve with them) often don't know how to write to an audience that isn't the five people that they collaborate with, that already know their subject really well.

I think scientists should spend more time figuring out, not necessarily how to be persuasive, but how to be interesting. I think a lot of scientists are oblivious to how to keep someone's interest in a narrative fashion. And even though it's scientific writing, there needs to be, not necessarily drama, but pay-off.

OK, here's something I have learned that I forgot to mention. I had this revelation a couple of months ago: I write in a dramatic style and it's contrary to what people might be looking for in a scientific article. What I mean by that is I tend to build up the tension in a question, which means that an introduction of mine will start vague and unfocused and the last sentence will be "ba *bam*!" This is not necessarily the way to write a scientific paper. I'm remembering back to a humanities writing course where you have topic sentences in each paragraph. We were taught you should be able to go down a paper and read all the first sentences and that should give you the story. And then you read the rest of the paragraph if you are interested in more detail about any one of those. What I tend to do is the opposite: I end the paragraph with the topic. I keep on getting more and more focused, and at the end you figure out what I meant.

But I'm finding that that's killing me. I had this discussion with my advisor on a specific paper because we were finding that the best ideas were always buried. As opposed to the best ideas leading off something—leading off a section, leading off a paragraph—they were always buried where someone might not find them if they were just skimming through a paper. I think it's not helping the scientific style of my papers in that I know scientists want to quickly scan a paper and find out what the points are. The first sentence should just be a bold declarative: "we're going to do *this*" instead of having that be at the end where you've had to wade through everything that I've said. I think my personality's more comfortable with the persuasive approach. Coming from copywriting and creative writing, I have a great love of fiction, and I just love writing that is alive and that has flow to it and thrust and dramatic tension and even, you know, gags. Maybe there's no place for that in scientific writing—well, there's less of a place.

GAO BOLE

Walking into Gao's office, I'm assailed by colour and texture and light, and the almost life-size figures in the corner seem to be taking notes. Meanwhile, Gao is full of welcoming energy. He's made an international career out of his remarkable track-record as an award-winning teacher (his fields are chemistry and education)—he is always in demand and travels extensively (which, he explains, is how he's acquired most of the treasures that fill his office). I have chosen to place his narrative in a chapter on the writing community because of his unique take on writing as an oral endeavour, undertaken within a wider community, and the impact on his writing of a group of "trusted people."

THE IDEA THAT SOMEBODY IS TAKING SERIOUSLY THE JOB OF READING SOMETHING THAT YOU'VE WRITTEN . . . IS JUST INVALUABLE TO ME

Back in the days when there were actually bookstores one could go to, one could go to a science section of bookstores and find lots of books on mathematics and theoretical physics and so forth for the public. And chemistry has always been really under-represented in that area. There was a handful of people, Oliver Sacks I think has done well. There's a chemist at Cornell who's tried, who thinks about these things pretty broadly, and a couple of other people who do write with that intent in mind. But it never took up more than an 8- or a 10-inch area of the bookstores. Why hasn't chemistry ever generated its Carl Sagan? I don't know the answer to that question, really. You've got terrific people who can talk about black holes for God's sake—you'd think somebody could talk about drugs and write this kind of business.

I think what science writing does for the public is to give people a deeper understanding of something they thought they already knew about, but didn't know deeply enough, but really didn't even understand they could understand it. And then all of a sudden you feel like you're just tapping into that whole "Secrets of the Universe" thing. As a student that was one of the things that appealed to me about chemistry; it wasn't the classes, it wasn't the experiments—it was that at the end of the year I really felt like I was learning something, and I was learning something that really matched up with things in the world that I knew about.

It really comes down to one question—can you tell a story? I think writing or telling anything is about a story. A good story just means that after the first sentence you haven't tuned out; you're actually interested in hearing what comes next. And it's up to the writer or speaker: that's all under your control. What do you say in order to try to whet the interest and desire of the person who's reading or listening at the time? I think you have to do that in class, every day, every moment.

I have been on the side of justice and good taste when it comes to this question of online learning. Which means that I think they're all bullshit. I really do know in my heart of hearts that you and I sitting here face to face having made a social commitment at this moment to be thinking about this is simply different than us doing this by Skype, by phone, by non-synchronous methods. And I think whether it's me and 10 students in class or me and 400, the commitment that people make at that moment has a special character to it. There's a different kind of social commitment that just is not matched in any other way. So the process of formulating and narrowing down those ideas I really do see as a rehearsing of material. I do see the analogies to comedians and actors very much so, but especially I think comedians' material, as I want to hear how my idea plays for the public.

So the normal part of an academic's life is to be out on the circuit giving plenaries, giving keynotes, and I always feel like I'm testing out the material to see how it sounds. And sometimes it starts as a footnote, and sometimes it starts as a comment, and sometimes it starts as an answer to a question. But I always see myself as playing with the ideas. At some point along the way you feel like you've now got a coherent story that you not only think is defendable but one that you *have* defended, one that has had people jabbering back and forth about it and one that has changed substantially through that process.

If you were writing science, this would apply—absolutely. The story you tell absolutely depends on how it's perceived and reacted to by the people you're trying it out with. And it doesn't mean that the scientific facts change, right. Evidence is evidence. But claims come from a warrant that the evidence provides. And I can take evidence and tell all kinds of stories about it, and I think we do that in science quite explicitly, all the time. In my role as a journal editor or reviewer—I'll make up a number—I think that at least 75% of the time the criticism that I make that recurs and recurs and recurs is people who have over-interpreted their data because they wanted a certain story to be true and they haven't been critical about what the data tells.

Chapter 4

I think that's true in all of the things that you write and think. It's an enormous benefit to this profession that so much of what we do has an oral component to it; that there's a venue that allows us multiple times to be trying out our arguments whether it's with our research groups, whether it's in the hallway with your friends. One of my colleagues will be the first to tell you the reason we still interact is that when he's writing something weird that he thinks nobody else will look at, he will send it to me because he'll know I'll read it and give him the feedback.

There's a benefit that doesn't necessarily appear in all parts of the academy. I think of historians as lonely people who aren't necessarily surrounded by as many people and the opportunities that I have to work out their ideas as they're going on. Recently, the editor for a journal I write for saw me at a meeting and he said I'm inviting people to be guest editors; do you have something you want to talk about? And all the thinking I've been doing about a particular issue crystalized in the form of that editorial. But that editorial has its antecedents in talks that were being given two years earlier. You could find the ideas in the drafts of either my PowerPoints or talks, and I still keep notes on my computer. You could see all the little pieces that ultimately came together in a story that looks pretty good now, that people like, was hardly identifiable as a single story two years earlier.

I think you're always being persuasive. I think you have a point of view and you're trying to get people to understand what your point of view is—no matter what—even if it's self-evident. I think it's always true.

I find easiest the writing that has followed the greatest degree of planning and practice. I don't think it's of a particular kind but if you really have jelled the story, then I just think it comes rolling out of your mouth or your fingers. And I think the times that it doesn't are when it still needs to bake a little bit. Now sometimes that process does take place through the writing, obviously. It's just as useful to put it down and think about ordering it as it is to go out and practice with people. So that definitely happens too.

I definitely make discoveries, new ideas while I'm writing. I think at that moment you're just holding all those pieces in a different way than you might have held them, they're touching each other in different ways that they weren't touching before. And if you're at least dedicated to making your story coherent you'll find connections that you didn't find before. I think that's true for everything but I do think that the process itself is a part of science.

In this interdisciplinary space where science hits education there just are not that many people who live there and write there, and so the word that's probably used thoughtfully about things that I've done is *translator*. I'm translating the science into the non-science area or translating the work in education into the science area so that these people can make sense of it.

And the short answer to your question about why I ended up doing this is that I learned in my graduate programme that you're supposed to identify hard problems to work on, and you're supposed to take your creativity and solve hard problems that other people aren't working on. It was clear as a bell to me that working at the interface of chemistry and education was a really hard problem. And I thought I was well suited to work on it.

I think chemistry's notorious for being very conservative; after writing in the passive voice for 30 years, people begin to just think and speak that way. I attribute it to this narrow, narrow genre with fixed rules. All you have to do is copy. And during the most formative time while you're a grad student your writing is going through a very strong editorial process with your advisor and certainly the classical written thesis is totally dominated by this model that just gets propagated like crazy. I bet if you did a writing analysis across every single one of those journals it's a tightly conserved kind of thing. One of the articles that I pick up and throw out to people all the time is Swan and Gopen.[8] I adore Swan and Gopen. I really love what that article does. And I really feel that the most important thing about that article is how accessible it is to the audience and what it models. So the notion of writing a review and making sure you show examples and take people through the alternatives and all that kind of stuff, God I've pulled that out to show to people so many times when the thing that they were trying to do lacked those features.

I loved writing as a child. I loved writing; I was just no good at it. In college, I am sure we wrote lab reports which were just formulaic. I learnt to write science the only way possible: working with great people. So I learned the conventions. I certainly learned the conventions of the writing of laboratory science the way everybody else does. But I think that writing's horrible. I don't even consider that good writing. It's not a kind of writing I would ever do outside of that context, but I understand in that context you have to write that way. I could probably find in that closet right now the very first time I ever tried to write something that was about this business—pages and pages of yellow paper with horrible sentences and no coherence whatsoever. It was a stream of consciousness. And one of my colleagues here in the department was an awesome writer; a really great writer from a technical standpoint. I would not have considered her a poet by any stretch of the imagination, but that woman knew sentence structure and she had a sense of notions of coherence and how to order stuff. And fortunately she was a dear friend and very open in her mentorship which is to bleed red on anything that I gave her.

And I have a wonderful peer group. You know, you either come to value editors or you don't. But the idea that somebody is taking seriously the job of reading something that you've written and trying to tell you if you're saying something is just invaluable to me. So anybody who will read what I've written and give me feedback is just wonderful, I think.

So while I may not have emerged out of a context where writing and issues were talked about, I'm certainly in one now. Absolutely. Whether it's the broader national community of those people I connect with—but certainly in the day to day works as in my department—absolutely. I like to think I've contributed to that because I have these other great people I've worked with and so the younger people I've worked with have just been naturally drawn into that. And they don't come easily the first couple of times, but boy by the second or third time they absolutely see the value. So, the people who have influenced me? You know that, that subset of trusted people. That really, really sincere group of people that was interested in my career, in my development, that took the time to just read this horrible stuff and respond to it.

THE WRITING GROUP

Elizabeth and Sally are friends who work in related fields and who, inspired by a writing seminar, set out to make a difference to their own writing and the writing of others. They're both mid-career scientists—Elizabeth is more senior than Sally—and they're under pressure to get publications out. The writing group doesn't turn out how they expected, as "why aren't we writing?" dominates successive sessions. But this writing group certainly meets Bereiter and Scardamalia's definition of a learning community, as the facilitators lead out of their own lack of knowledge, and the team develop their understanding together.

Elizabeth: I'm writing more. Writing a hell of a lot more. We both went to a writing seminar last year. And we both came away going "Wow! That was so cool. That was so good." Because we were both struggling with getting our writing done; there just wasn't enough time in the day. She was so inspirational. And Sally said "we should start a writing group."

Sally: And I think the thing she also did was she blew away all of the excuses we have always used. I wanted her to help but I was also thinking, "oh I can't do this because of this, this, and this" and she confronted all of those things and by the end of it, I was like, "oh, okay then—I have no choice!"

First, for scientists it's important that she gave some evidence that the approach she was suggesting actually resulted in changes. Measurable, quantifiable changes. And that for us is important 'cause we're not the kind of people who will take things just on face value. And then the other one was about the "dispositional fallacy" which is basically "I'm not the kind of person that does things every day." And I've been using that one for ages: "Oh I need to set aside a whole day, a day or two days, because I'm just not the kind of person that does that." She's like: "Rubbish, everybody is the kind of person that can do that."

So with our writing group, we invited students and faculty to come but there was also an element of "it will be really good for us." I needed some extra motivation and some—what's the word?—accountability.

Elizabeth: The seminar leader said "Start a writing group. Start or join a writing group. Regular meetings with a group of colleagues can provide you with motivation, feedback, and camaraderie." At that time my students were just getting to the point where they were about to start writing, and so I hadn't really encountered this as much of an issue whereas Sally already had. She was seeing it for herself and also the people she was advising. And so that's what we canvassed around just within our biological sciences group. Because I suppose we have how many—eight, ten, twelve postgraduate students, something like that? We said, "who wants to come along to an inaugural writing group to see if this might be something that's helpful" and we had the first meeting.

Sally: We had the first meeting in December because we didn't want to wait until after Christmas to get the motivation going. I think we started off by explaining to them why we were trying to do this and what some of the objectives were and how we foresaw the meetings working. But we also made it really obvious to them that it wasn't us leading the group and we weren't going to be giving them answers or telling them how to do things; it's going to be as a group of people trying to figure it out together.

Elizabeth: We've facilitated the group. It has fluctuated, you know sometimes they just come in and they just sit and look at you. And I think that we've usually discussed an agenda before the meeting about what we want to cover. I would say this has been a very organic process. One of the things we kept ending up coming back to (which would keep deferring the planned agenda) was that people were still not writing. And it's this problem of not writing which has been the major one—we spent parts of the first six meetings trying to get over this. And I think because we were struggling it might have been actually quite helpful to the students to see that, oh, we are all in this together.

And in the very first meeting we sat down with the people and said, "well, why aren't you writing, what are the problems you find with respect to writing, why don't you get started?" and there were a number of main points which were brought up. The first one was that they were overwhelmed and they had no idea where to start. So these are students who are largely doing theses, and if you're not experienced with writing a paper you don't know where to start. It's this huge big unopened box and to even peek under the lid was quite a challenge.

The second thing was they didn't know what they should be writing about, so they didn't know what they should be addressing, they didn't understand the

Chapter 4

question they should be writing about. So "here you are, postgraduate student, here's a thesis, write a thesis." The third problem they said was they didn't know how to express the material.

So the first thing that we really got into was goal-setting. We keep going back to the comment: write every day, write every day, write every day. And to set the goal of a) doing something every day, and b) having a specific goal that you're heading for in that writing period. And we discussed a number of different goals. We probably discussed writing a paragraph or writing two hundred words. And I think one of the best things that worked for me was when one of the postgraduate students said "Why don't we make a goal based on time rather than on work." And that's been really helpful to me because some days I can get through a lot of work, some days I get through a very small amount. And one of the things I realised part way through is that I set myself goals that are too big and I become despondent about it.

So the setting of a realistic goal is a major component to getting people to start writing. I know I can do an hour a day, or an hour and a half a day, rather than I want to get section four finished by the end of the week. Because section four might actually take me two weeks.

In the first meeting we were going to talk about why people couldn't do it and then we were setting little exercises to say "well here's a little section, have a look at this and see what you think about the quality of the writing." It was a completely inappropriate exercise for the time.

I think at the second meeting we used one of the exercises that the workshop leader used in the seminar. To overcome the belief that you don't think you can write, she gave us ten minutes of free-writing time. She said "for the next ten minutes, you just write. And I don't care what you write about." And we did use that as a technique and I don't know how effective that was but everyone covered quite a lot of writing.

That was up until Christmas. We set goals. We started off with a writing log fairly early because one of the things we talked about was having accountability, having a writing buddy. So Sally might email me at the beginning of the week and say "I'm going to write—this is my goal for the week." And I would email Sally and say "these are my goals for this week." So what we thought was that for all of the people that were coming along, we'd have what their goals were, and then at the next meeting they would check what they had done and then set new goals. And I think it was quite interesting—the complete failure of a number of people to achieve anything. So motivation was a problem and I think it still is.

We set goals for the Christmas period. After Christmas, how many people had done writing? Well there were heaps of excuses. Lots and lots of excuses for why people hadn't achieved what they were meant to be doing.

Sally: The seminar leader's suggestion was that your writing time should be just for writing, but we all struggled so much with that. We just don't have any other time to put aside apart from that two hours we've set aside . . .

Elizabeth: So we had to adapt her method to one that was more suitable to science things. So at the end of January, we're still talking about why people aren't writing. How we need to get people going. Then we actually got started on the second point. The first point was that they were overwhelmed, the second point was they didn't understand what the question is. And we realised that very few of them were undertaking research with a clearly defined hypothesis. They didn't know what a hypothesis was, and that's where Sally's experience came in.

Sally: The advisor wouldn't recognise that there was a problem until they got to the writing stage. So with the kind of work we do—we sit down, we brainstorm and we have a loose idea of the question we are trying to answer, but very few advisors, I think, sit down at that stage and say "right, write down your hypothesis." And maybe we should. Some of my co-advisors are reluctant to get students to do that because they don't want to constrain their thinking at that stage, they want them to keep their mind open. And then so they do the experiment, then they analyse the data, and then they get to the writing stage. By that point any action is remedial.

Elizabeth: I'm not sure that I completely agree with that. I think that you have a major hypothesis, then you set up one group of experiments, and there's nothing wrong with defining a sub-hypothesis at that point. And your hypothesis will change as you get into your research and you get this answer here which will alter this hypothesis over here and that will ultimately feed back to the primary hypothesis at the beginning. But I think I was quite surprised at the ignorance level—including me—and I was quite surprised when we actually sat down and looked at the nitty gritty of what should be driving the research.

So at about the beginning of February the size of our writing group had decreased down to about five or six regular attendees. What we did at the beginning of February (because we wanted to be able to look at this process), we got the people who attended to write, and put into a sealed envelope, their feelings about writing *before* they started the group. Because then at the end of it we can open this again, and they can have a look at what they felt before they started the writing group and how they feel now, so we can see if we've made an objective difference.

Sally: One thing we need to decide for the next iteration is whether we going to open this up to students again or is it going to be offered to faculty? The issues are different and when we first started we hadn't intended to do all this didactic stuff. We had really intended to spend a bit of time to confront the "why aren't you writing" and then get into a stable state where we met regularly and workshopped people's material and supported each other in an ongoing way. And it's not that it hasn't been extremely valuable, but it hasn't been what I first envisaged that it would be. So we haven't really come to any decisions about what the next iteration would look like.

Elizabeth: It's giving them the tools. It's giving them the motivation. And tools and motivation is about understanding. And not being afraid to lift the lid of the box because "oh, yep, it's a bit of a jumble, but I have the tools to sort it out. I have the tools to organise it."

So we never got on to what we planned for it to be, right at the very first instance, which was people bringing their work and getting constructive feedback on it. We're getting to that now in terms of structure—will we ever get to it in terms of writing style? I don't know. I don't know whether that's so important. Because I think if you've got good structure a lot of the basics in writing style will follow. If you've got good structure you don't repeat yourself, you don't tangle yourself up in knots. It's all laid out there and you can sit down and, instead of faffing for half an hour and thinking "what the hell am I meant to be writing?" you think "oh that's where I left off yesterday, bang."

ELIZABETH

Elizabeth's office is chaotic. There are books and papers all over the floor, and throughout the interview people are putting their heads round the door asking for advice on a variety of topics. Elizabeth speaks to me, as she speaks to these intruders, with focused, impatient energy. She barely pauses for breath between ideas. I'm struck by her generosity in sharing ideas and time—with me and with her writing group; she's thought a lot about writing, perhaps because she writes to such a wide range of audiences in such varying genres. As well as writing science (books, journals, teaching materials), she writes fiction for young adults, and newspaper articles for young children. She describes her teaching and research as closely interconnected, each feeding the other.

THE ONLY OTHER THING I WOULD SAY IS I AM PASSIONATE ABOUT WRITING

The mechanics of how I write has changed a lot since that first writing seminar. I used to try and do it sometime during the day, which doesn't work. But after we were talking about putting a dedicated time aside, I started trying it at seven in the morning. So the mechanics of it are that I'm in here usually by seven, or seven-thirty on the darker mornings. But I will get two hours in at the beginning of the day when it's quiet, and because I'm doing it every day I know exactly where it is I've left off the day before and I don't have to think "where the hell was I?" I come in, I make a cup of tea. I usually have my breakfast next to me and I beaver away. It depends what stage I'm at.

So the book concept that these notes will contribute to started about four or five years ago with a colleague, before I left Scotland. We've really only been working on it probably in the last 18 months and F was the one who identified there was a need for this book. One of the problems with the topic is it scares a lot of people off because it's got a lot of terminology, so all the books that are out there at the moment are not user friendly for somebody who just wants to get some basic information. So we're going to do a Noddy's guide. We've come up with some new concepts about how we're going to present the information. So that was done—brainstorming it really—initially just with R because he is here, and then a telephone conversation with F who's in Scotland. And then I've been the one who has largely taken it forward because I've had to write these notes for another course (in Europe) anyway.

We had drafted out a pretty good outline for the book proposal that we had worked up together and we had worked out who would write out each section and roughly how many pages and how many words were going to be involved in each section, and this was also required by the publisher for the contract. But what I found was as I was writing up the notes for the European course then I actually ended up covering the majority of all of the sections. So the first draft of the book is largely going to be all my writing, and my colleagues will come in and add in little bits, and F will end up probably being the major editor of the book.

I've been teaching in my field for about 25 years, but particularly in the last four years my approach has changed, and in particular it's changed in the last 18 months when we came up with this new concept which is part of the book design. I've been changing the way I've been delivering the information to both the veterinary students and the science students that I teach, so there's a lot of interplay between having got this outline and the new concept, me thinking about it, me putting it into practice by delivering it to the students, getting their feedback, and then realising where the difficulties in their comprehension might be, because as you know the more you deliver information the more you understand it. If you have been delivering the same or similar information for a number of years, you'll gradually distill more and more the essence of what you're teaching, so you get better at wording it and the concept is more readily picked up by other students. So those notes, plus the delivery to the students, reshaped a lot of the information in my brain which led to the notes for the European course, which were also written in a format so that they could form a large body of work for the book.

Chapter 4

So I had the foundations, I had a reasonable understanding of the structural arrangement. Initially I picked up those notes and I put them down as my format and then rewrote from there. As I repackaged it I realised I needed to understand a lot more and that involved going out into a number of other texts. I sat at my desk and some days I'd have half of that bookshelf on this table, drawing from that and that and that, and checking and cross-checking. I'm looking at it from a new perspective, having to say to myself "if I look at it from that perspective I wonder what such-and-such and such-and-such say about it?" You know how you read a passage with one bias, or you're seeking one bit of information. Then if you change the way you're viewing the information and what you want to get out of the passage, you can read the passage and get quite a different perspective.

I do creative writing, that's my weekends. For the book I'm writing now I have a rough outline so it's plotted out from go to whoa, and when I sit down I've got an overview in my head of where the story is going. But then I've got it down in sections as well. I'll know the section and what's going to happen in it. I begin by reviewing the last section that I wrote, read over it, edit a little bit, then I'll write my new section. So I know what happened before and I know what's coming up next. When I write it pretty well comes out—not fully formed, it needs editing, but I'll write it out in full prose.

To go back to science, I started writing a new manuscript earlier this semester and it was in an area that I really didn't know. So having done the database search and having pulled out all the papers I thought were relevant, I just started working though those papers. And as I read through a paper highlighting the bits that were going to be relevant to the paper—I had a rough idea of the outline of the paper and so could say "that bit goes in this section, that bit goes in that section" so by the end of reading the references the sections would be populated with information from ten references which were all linked but needed then to be turned into a synthesised whole. So they were bullet points, and then I might find that Joe Bloggs and whosey-whatsit had written a similar sort of thing, so then I could take those two bits of information and compile them into one sentence. That's a slow and relatively painful process. I don't know another way to speed that up.

Much of the writing of the materials and methods of a paper will come from previous papers. The writing of the results can be somewhat slow because you've got to go back to your data and have another look at that. And then the synthesis, the discussion, when you're looking at your results compared to what else is out there, much of that information has come out of that first draft of the references. Because this was such a new paper and a new area, I didn't have a really good outline for the manuscript because I didn't know what I was going to come across until I started reading.

I don't start writing with the material and methods. I start with reading the literature. I mean, in an ideal world (and this is something I've really only appreciated in the last couple of years) writing a paper should be done at the same time as you're doing the experiments. So, you can't write the results, obviously, but if you are going through the literature at the same time as you're doing your experiments, then the literature will provide you with the information you need to complete the tests as you planned, but also provide you with thoughts on what else you might need to plan experimentally to go into that paper. And if you are attaining that knowledge through the reading at the same time as you are running your experiments, it means you don't get to the end and start writing up your experiments and think, "oh, blow, I should've also done this and this, because now I could be challenged on that by the reviewers," or "my data's incomplete," or "I made a mistake." So I'd probably start with the literature. And when I get bored with doing that then I'll do the easiest section like the materials and methods, the mindless stuff.

I think one of the most interesting things about the writing group that we've started is that we've realised (not realised but clarified) that the whole structure of the paper is constructed around the hypothesis and the aims. So for example, let us say you have the hypothesis "if x then y." In writing the introduction, you have to introduce x, you have to introduce y, and you have to introduce the relationship between the two. Your hypothesis "if x then y" also determines your aims. The aims then automatically determine your materials and methods. And the aims then also automatically determine the laying out of the results. And the relationship between x and y automatically determines the structure of the discussion. And that has just been a fascinating realisation. It's so cool. We were working through that about the same time I was starting to write this paper. And I think the next time I write a paper I will be starting to write it as soon as the hypothesis is clear.

I think I have a much clearer idea now about writing papers. My old mentor in the UK would always write his papers as he was doing the research. And he would say that to me and I never really got my head round that—I was always too busy at the bench. But I really see the wisdom of that now.

You might change your aims as you go along, depending on your results. And that is why you should be reading as you go. Because then you are going to be much more aware of that potential. Yes, we do throw up surprising results, and that's fabulous. But that doesn't change the structure of the paper. You've got to be prepared for yes or no or maybe, depending on the situation, or for "bugger me, I haven't accounted for that variable." That might also lead you to the end of that paper and saying, at the end of that paper "our hypothesis was if x then y, however we have found blah, blah, blah, this leads us to our new research

Chapter 4

which leads us to our new paper." So it's not black and white. It is an organic process. But I think it's going to be a far clearer process of development if you're writing as you go along, because that gets you thinking and distilling. You distill the hypotheses or you refute it, or you change it.

That's my opinion, my thoughts. And they have changed hugely in the past six months because of the writing group. Sally and I were laughing about this—who are we to run a writing group? Okay we've got some experience but, by golly, my ability to write papers—I love writing but I hate writing papers. Why? Because nobody ever taught me how to write papers. My old mentor probably did try, but I was probably, fingers in ears, "I don't want to learn that you've got to write them as you go along." So for Sally and I to say "we're going to set up a writing group" which conveys the thought that these people must know what they're talking about, well, actually we don't. But that's what's been so wonderful—we've all worked together, and together we've come up with this concept that if you've got your hypothesis clearly defined, it's actually pretty easy to understand what you put in the introduction and how you approach the discussion.

I've learnt so much from that writing seminar because she hit the nail on the head about why the writing ain't happening. Our careers depend on us writing. And you know what? The writing of the research papers is, unfortunately, the thing that is constantly put on the backburner. And why? Because teaching has absolute deadlines. Nobody's really standing over you to get that paper out. Your career depends on it, but no one's really standing over you to do that.

I think if you're not aware of your audience, you're barking up the wrong tree. Absolutely. End point. Begin with your end in mind. The sort of work I do is going to have a much wider audience because it pertains to cell culture, it pertains to neuroscience, it pertains to specific neurological diseases, so you can take your pick out of those. That could be thousands of people. I write from school kids to scientists. So a fairly wide range, and in-between that's undergraduate and postgraduate and professors and post-docs. And so of course my style changes depending on who I'm speaking to. But one constant feature I aim for is clarity.

And the second thing which is a common thread from an 8-year-old to an 80-year-old professor is to try and think, what is their experience and perspective? And what you're trying to do when you're writing a paper is to find the best way to pass on the information—it's another form of teaching. I'm a teacher. That's my number one, I think. To help people assimilate information, you've got to think, well what hanging hook for this new knowledge have they already got in their brain? Most hanging hooks are shaped by experience and knowledge at that time. So for an 8-year-old—I was just writing a piece for them last night—their world is small. An 8-year-old is: "this is me, and here's my mum and dad, and there's my dog, and there's my school, and this is my experience, and my experience is very much about me." So what I was writing on last night was a reply to a student who had asked why birds take dust baths. And I was going to go on about how dust baths are useful for removing parasites, but I needed to introduce the concept of why they got parasites in their nice, warm, dark feathery places. And so, what's a parasite? I said "imagine what it must be like to have creepy-crawlies crawling around in your hair." So I'm trying to link in to their level of experience. I'm not saying I'm doing a good job of it, but I'm trying to think how an eight-year-old would think.

Whereas when I'm writing for a scientific audience—and undergraduates are different from postgraduates, who are different from a research colleague—I'm going to assume a level of knowledge, and that they're busy people, and I'm going to assume that they will want clarity, and they will want to be able to skim it. So I will tend to use a top-crust style of writing, which is: I'm going to tell you in my first sentence or my first couple of words what this paragraph is going to be about. If I'm writing for somewhere in between, like an undergraduate who's got a degree of knowledge, then I'm going to keep the terminology from overwhelming the concept, and I'm going to be trying to pull out the concept—the concept is the number one thing I want them to get, the terminology is number two. So I have a priority of how I want you to pick up this information.

There is a huge role for metaphor in explaining new concepts—it goes back to what I said about the 8- and the 80-year-old: finding the hooks of knowledge that they already have, using those, bringing those forward to say "okay, you have an understanding about this already, this is actually quite similar to this." If you are offered a new piece of information and you can immediately put it into your filing system or hang it onto one of those hooks in your mind, then you'll retain that information. But if I offer you a completely new piece of information for which you have no reference point, no hanging hook, it's like opening the hall closet. In there will be hanging hooks with specific functions. I'm going to put that on that hook, that on that hook, and that on that hook—hang on, here's something I ain't got no hook for, I'll just throw it in there. Outcome? It's going to get lost.

So I think a key concept in teaching—and this is what I believe communication is, be it written, verbal, oral, science, creative, whatever—is to ask who is my audience? What knowledge hooks have they got? How can I link this new information to those hooks?

We could do things differently. I think there is a huge need in the university for people to mentor others in writing, I think there's a need within the university for people who are in mentoring positions to know how to give feedback as to why that section didn't work, why they've changed the writing. I mean, I can change someone's writing around but I can't necessarily explain to the person why I've done that. Just "because it reads better like that." I think that the willingness is there, the intelligence is there to do the research, but writing it up is the challenge. We could have writing groups; we could teach them how to write.

The only other thing I would say is I am passionate about writing.

CHAPTER 5
THE DEVELOPMENT OF THE SCIENTIFIC WRITER

> I think you become savvier about not just being a good writer but at writing to an audience. . . . And I think that's where I'm at. I know who these people are who will keep writing the same papers to the day they die, the same sort of formulaic kind of stuff. I want to get savvy and become more proactive than reactive.
>
> — Lizzie

One of the questions of this book, as outlined in Chapter 1, is whether the writing of scientists changes over time post-Ph.D. Do scientists' writing activities change and broaden, and do their beliefs and attitudes to writing change with them—or, perhaps, do their beliefs and attitudes cause them to engage with new writing tasks?

The findings of this study in relation to these questions were somewhat equivocal: while almost all scientists experienced a change in writing activity post-Ph.D. (from a primary focus on writing their own research to supporting the writing of others—see Chapter 7), the extent to which the audience for their work broadened, and their attitudes and beliefs changed over time, was more variable. Some defined a narrow field, addressed by a specific hierarchy of journals and discipline-specific organisations, which they wrote for and engaged with. One of the participants who worked in this way, when asked about whether he thought about his audience when he was writing, commented that he did indeed, since he knew all of them personally. They were 8–12 scientists in his field who met together regularly at conferences around the world, wrote together, engaged in lab rivalries, and reviewed each other's work. These are the routine expert science writers, who define a field and work narrowly and extremely competently within it. One of these scientists submitted his work primarily to three specific journals (for one of which he was co-editor), commenting that he had never had a paper rejected "because I know my stuff." Retaining this highly specialised focus was an individual decision, not driven by field (one participant, for example, explained how his work could be adapted to a more interdisciplinary context or have been appropriate to public interest) or outside pressures—indeed, could be seen as being maintained *despite* external pressures to engage with what one participant referred to as "big science" and its associated funding.

Chapter 5

On the whole, those scientists who engaged with broader, more diverse activities as they developed seniority, tended to be critical of these narrowly focused scientists—this was certainly not a path they wished to take. And yet this group of narrowly focused, highly specialised scientists included individuals who were extremely successful, even award-laden. Not engaging with "big science" did not seem necessarily to inhibit their careers or, perhaps more importantly, the way *they* wanted to grow their careers. I heard no regrets.

The scientists whose writing activities changed over time tended to express different beliefs about the purpose of science and more complex motivations about writing science, which led to individuals seeking out interdisciplinary research partners or opportunities to engage with the media or social media. While these individuals maintained a strong interest in moving their field forward, and most were engaged in writing their own research, they saw and pursued opportunities to broaden their focus.

In this chapter, I have chosen four interviews that illustrate the progression from narrowly to broadly focused writing, and the beliefs and attitudes associated with this shift. Grace, a young post-doc, is engaged in writing in a narrow field. At the time of the interview, she lacks confidence as a writer, sees writing and science as separate activities, doesn't see writing as persuasive, struggles with issues of audience, and relies on imitating her advisors' writing style to develop her writing. Yet I have chosen her narrative because it contains the seeds of a growing understanding: she enjoys writing, is developing resilience, sees the value of adapting to feedback, and is taking steps to broaden the audience she engages with.

Lizzie and Paddy are at a different stage in their careers, and both show an understanding of where to go next to develop themselves as scientists and writers of science. Neither is content with a narrow field. Paddy is about to begin a research project with a group of writing researchers, and is considering how to engage with a public audience. Lizzie describes herself as being on the cusp of the next big step. These two narratives demonstrate more sophisticated approaches to science and science writing: they talk in complex ways about audience, persuasion, process and style, and they enjoy writing in a range of contexts.

The final narrative in this chapter comes from Lemrol, someone who has reached the last stage of his career, and who exemplifies the adaptive scientific writer. Like Richard and James in Chapter 2, and Catalizador in Chapter 6, his interests generally, and more specifically in writing, stretch well beyond a narrow discipline. He is the master of his craft—highly resilient, strongly innovative, endlessly curious. His research, and his writing, is now influential in a range of contexts, and he has a significant role in shaping the next generation of scientists. He is the model of the adaptive end-of-career science writer.

GRACE

Grace was the only person I interviewed in a lab, in her white lab coat, surrounded by equipment. She's somewhat distracted, not quite sure why she agreed to this interview. Perhaps in keeping with her status—she is early in her first postdoc (a three-year project) in the field of marine science—she has the simplest attitude to scientific writing. In her view, writing is not part of science, it doesn't have to be persuasive, and her approach to style focuses on mimicking the style of her advisors. She has yet to develop her own style or a sense of ownership of her field. But one thing of interest is her description of writing a paper as an organic process of writing all the sections almost simultaneously.

I Don't Think Writing Is Part of Science

Oh goodness. Shall I describe my project? OK, well I did some preliminary experiments—this is before starting to write it up—and found that I had discrepancies with other papers, and then we decided to take it forward and do it as a project in itself. So I think then you, or we (it's hard to describe how it goes) start with the introduction.

But also at the same time we're looking at the results, so I find that I tend to do both of those things at the same time: you know which direction you're going in and also you don't want to tread on other people's toes, so if someone's done the work before you, you don't want to repeat what they've done. I'd say definitely the introduction and the results at the same time, but the results are ever-going until you've finished your experiments. And then probably the methods as well, you're starting to write those up as you're doing them so you don't forget them. And then as you're getting your results, you're formulating an idea of where you want to take the discussion. So I would say the discussion would be next, and then finally the abstract.

I'm not very good at seeing a big picture. I get very bogged down in details, so it's good to have the different paragraph headings. I try to keep to those headings and then you can formulate the plan and see how the paper flows or how your write-up is flowing. I do my processing on paper. I print out many, many copies and keep going through it. Even after a day's editing or changing, I'll take it home to read on the train and sometimes I can't quite believe what I've written because it doesn't make sense. So it's obviously been a brain dump from my brain to the paper.

When I've got it to a stage where it's all written up, I pass it on to my advisor. He's very good at making things concise. So what I say in two sentences he will say in one sentence, and I don't know whether that's because of his experience, or just that he's very good at writing. But you learn from that. Recently we've come to an agreement where he'll say "that paragraph needs reducing by half" or "that section needs to be reduced, and the lists that you've done there are too long so take out some of the detail and put them in the paragraph afterwards." So he'll send me away to do that myself which is really good—rather than him doing it and then me learning it parrot fashion. I think I'm quite quick to learn, so once I've seen how a section has been corrected, I would then absorb that correction and apply it to the rest of my writing.

We do have other authors but generally, within this post-doc, it's mainly my main advisor here that would do any editing; the other one lives elsewhere. While he might pick up spelling mistakes, he doesn't really change that much, and I don't know whether that's because he thinks it's fine or because he hasn't got time to go through and change all these things. Sometimes I pass on my papers to family members just to read through—I know that some of the science is probably a bit gobble-de-gook for them, but as long as they can get the general gist, I see that as a positive thing. My grandmother loves reading through the papers because then she gets to understand what I'm doing.

Whether I think about my audience depends what I'm writing. If I'm writing a scientific paper then I don't really think about other scientists—you just sort of write—I never think about them. Whereas if I have the public in mind, then I would definitely think about the audience; and that's probably also when I would pass it to a family member and say "do you understand this?" because essentially they are a lay person and the kind of person that would read it.

I find writing for the public more enjoyable than writing for scientists and possibly easiest as well. I find scientific writing more of a challenge. But I want to get better. Self-improvement is what motivates me. And I do enjoy writing, even though it's a challenge sometimes.

I don't think scientific writing should be persuasive. I think sometimes it can be; it depends on the writer. I think if you're clever and you're good at writing then you can probably be very persuasive. But I don't think I'm that good at writing—I'm bad enough at just writing up my results to get published, let alone to publish it with an intent to persuade people.

I think if you are able to adapt then you can survive. Like we've had one paper—the one that I'm working on now—rejected by one journal and we've had to go back, rewrite it, do some more experiments and submit to a different journal. We've had to change our stance and the way that we've written it. I think the first draft that we submitted to the journal, was too . . . well, it's criticising entrenched methods. I think that upset some of the reviewers. So with the second submission, it's a lot looser; we showed that there are discrepancies with the method, but we're not so forthright in saying that another scientists' methods were wrong.

Writing is very important. You've got to keep writing. It's an avenue to show the world what you're doing, what you're working on, what you've found out. But I don't think writing is part of the science. I think you do the science and then you write it up—I can't see how writing is part of the science.

PADDY MCCARTHY

I interview Paddy, a postdoctoral researcher in experimental freshwater ecology, outside on a cold day in an icy wind. He has no office of his own—just a lab shared with other young scientists who have chosen this day to be at work. The various places we've tried around the university have been too noisy for my recorder, so we sit outside the library, our hands turning slightly blue, and talk. What is significant about his discussion, from my perspective, is his determination to see writing not as a thing in itself, but as an integral, inseparable part of the research process. He can't really tell me how he learnt to write or how he teaches writing. Instead, his focus is on the entirety of the research process. He's not a writer, "just a scientist," but for him writing is part of being a scientist.

I'M NOT A WRITER. I'M JUST A SCIENTIST

When I was a child, I used to write short stories and little books. I write poetry now, so I still write for pleasure. I really enjoy scientific writing as well, so I guess that is writing for pleasure too. Most of the time. Writing is something I really enjoy. But I don't have enough time anymore and most of the reading that I do is journals and papers. There are so many books that I want to read and they're all sitting there half read.

I work primarily on my own, but there are quite a few people in my research group; there's one Ph.D. student who is directly linked to my project, so we've worked quite closely together. And my boss is very involved, hands-on in the project too. It's a lot of solo work but with a team around me when I need it as well.

The project I'm currently on is for a grant that my boss won. So in terms of designing a lot of the core ideas, that's already in place and certain boxes have to be checked over the course of a three-year project. But then there's a lot of leeway within that as well, so you can put your own stamp on it and contribute your own ideas. To give you an example, my most recent field trip involved setting up an experiment that I had designed with my leader's help. So a lot of it *is* coming up with your own research or getting involved with the rest of the group and helping them with their data analysis or writing. I've come into a project that has been going on for years, so I'm also contributing to, or writing papers for, work that I haven't actually carried out.

The way I usually start to write a paper, if it's a paper based on data, would be to start with the results section and get the story clear. So I'd look for the key patterns in the data, weave them together into a story that makes sense and that you can engage a scientific audience with, and then work backwards from there. I would write an introduction to that story next, which obviously has to make sense in the context of the results. So the various introductory paragraphs should talk about the background area to each element of the story. The methodology's obviously fairly set anyway because that's what you've done. And then in terms of discussion, I guess it's starting very narrow by summarising the key results that you've found and then going through each of those results in detail, putting them in the context of the wider field and then broadening it out more and more so that you can relate it to the work of others; highlighting how it's advancing the field, or what the new questions are, new gaps that we realise based on these results and what the next steps need to be. I guess that's pretty much the paper written.

A lot of the processing is done in my head. I'm not one of these people who actually writes down a plan and draws a nice schematic; but to me the story is very important, so I plan it in terms of a story. So I would need to identify what the key elements of the story are; these are the themes that I have to address in the introduction; these are the themes that will have to reappear in the methodology so that you can see how each of them was carried out, and then they are the themes that I need to discuss and interpret and develop in the discussion. It is very structured, there is a plan, but nothing really formal.

There's always room for improvement I think, no matter how good you get at writing. I really benefit from talking to people, or showing my writing to someone who will look at it from a different perspective. In my most recent paper, there were a lot of co-authors who took on that role. Some of those co-authors didn't even have a big involvement in the paper, so it was almost like getting an outside person to look at it. But normally I would send it to the next most leading author in the paper, get their big input on it, and they would probably be involved in a lot of the writing as it develops anyway. If it's a high profile paper, you'd want to send it to a couple of people—maybe outside your university—just to get their feedback. Sometimes I'd go to a colleague down the corridor and say "look I have this paper, do you mind perhaps taking a look at it?"

Since most of my writing is scientific, I find it quite difficult to communicate complex ideas to an audience that is probably going to get bored by the details but excited by the key themes and topics. Sometimes you've got to write a press release for a paper and that's challenging too, writing in a very succinct way, selling the story to a general audience. And sometimes you almost start bashing yourself as well, because you get so caught up in your little bubble world of scientific journals and other researchers that you forget the more hands-on applicability of your research in terms of conservation or management perspectives. And it can sometimes be an eye-opener when you have to say what the real importance of your research is in one of those journals.

On this recent field trip I did very little desk work; you get into research mode and it's long days in the field. In experimental freshwater ecology you could be out most of the day, and it's very intensive work that tires you out, so the last thing you'd want to be doing is writing at the end of it. You make all the plans before you go there so you know what it is that you're doing; but at that stage you just want to set up experiments, carry out your survey work, collect your data. You're not worried about how it's going to fit into writing these different papers. You almost have to be a little bit distant from the end product because it might influence how you carry out your research. You just really have to do it in your logical, scientific fashion, collect it all as best you can, and then just trust that, later on in the lab, when you're processing samples, the story will start to emerge and then you can start writing things down.

I guess scientific writing is persuasive. You have your questions that you want to test; you probably have your idea of what the answer's going to be, and it is exciting when the results confirm your expectations. Then you want to persuade the reader that this nice piece of work that you've done was well thought out, was well executed; that the results that you're presenting to them are believable, full of integrity. You want to persuade them that the results, which you're saying have all these characteristics, are going to be really interesting to them and will forward our understanding of some particular topic.

In terms of the writing I like best, it's definitely forming the story, trying to see the pattern in the data. I don't know if that really counts as writing because a lot of that is storytelling or analysing or interpreting. I guess what's very rewarding is when you can start trawling through the literature of those buried studies that you haven't ever read before or that you didn't know about, and then you start saying "oh, somebody else had done something that proves an element of what I'm showing here" or "well that was a surprising thing but I can see why mine might differ to that." That's an exciting part of writing, still a kind of a learning process, seeing these other studies that relate to your work, even if it's slightly tangential.

I always find starting to write is the hardest part of it and it's the bit that takes me the longest; I can literally be sat for weeks just staring at a blank screen or just thinking "no" and going off and doing some other task. I think subconsciously you need that time to be able to process the story, the patterns that are there; and even though you're not actively thinking about it all the time, having that long lead-in period to writing somehow gets things gestating within you. It becomes a lot easier then, once you get into the flow, to really start, and everything kind of runs together. But getting it going, that's frustrating at times.

What motivates me with my writing? I guess it's a couple of things. One of the more facile ones is, you know, the same way as when you're a kid trying to collect stamps or whatever, and you want all the stamps or you want the best collection. And in one sense when you start getting into this publishing thing, you want to publish in better journals, you want to publish more papers, you want to collaborate with more people—so it's kind of "I want to do more" and you're going to get greedy. But from a more practical standpoint, I'm really interested in being the best I can be and having the best career I can, and a very important part of that, I realise, is building your CV. What you write is almost like your portfolio, and I guess the more research you do and the more varied the topics, and the more people that you do that with, shows that your research is valuable and of general interest. It's not just being a first author on the papers but also showing that you'd make a good mentor, that students you've done projects with can write really good papers as well. And I'm starting to like that process of not being the key person that's driving the writing but being there to offer a helping hand and seeing somebody else get to that end product stage. I'd say that's probably one of my main driving goals.

I think writing is part of science. It's not just there to communicate—of course it *is* there to communicate what you've done—but I think like what I was saying about when you're writing the discussion, there's an element of discovery to the writing as well. So just through the writing process—and I think every author will have their own unique way of doing this—you make connections between findings or the data that you've collected, the interpretation of that, and the work other people have done. I think only through writing do you make those discoveries and connections. And then reviewers might say "oh my God! How did you miss this?" or "you should have done that" or "have you read this paper?" It's all that process that makes the study not just the actual collecting of the data. It is this really nice integrated process of collection and then communication, but with feedback loops in between.

There definitely are different styles in different disciplines. Within the ecological sciences we don't have these definitive laws like they have in chemistry or in physics—but we're almost a little bit too hard on ourselves at times because of that. We take this really hard line on trying to be sure that everything is completely above board and as unbiased as possible. And we're very—I don't know what the word is—third-person voice, very cold and logical like "this was done" and "this is how it was done," but we won't tell you who it was done by because it needs to sound like a robotic process without any room for personal error. And I think that is because there is so much natural variability and confounding factors in ecology and, as a result, a lot of the elements around our research are kind of soft and open to criticism. There's no room for flowery language or anything like that—it all has to follow a highly logical scientific code.

Something that drives me is not just being pigeonholed into one particular field where you become the all-conquering knowledge master of that field—I don't think I could ever be one of those people. I'd rather be like the jack of all trades, you know, and have my finger in 10 different pies. I've changed my focal research area through different ecosystems throughout my career so far, and I've had collaborations with lots of people in different fields, whether it be empirical or theoretical, and I don't think I ever will be an expert in any of those. But I'd rather try and take the core set of skills I have and apply it to lots of different areas so that you're doing something that's new to a particular field and advancing that a little bit of the way, and then other people can go on and do with that what they will. So I'm trying to adapt what I've learned from one particular area to lots of other areas, and that's exciting, even though I'm never going to know everything there is to know about that new area. At least you've contributed to some sort of advancement of that field.

I thought I wanted to be a biochemist or a microbiologist when I started college, but on a second-year undergraduate field course we were taught by two people who were really passionate about ecology. I came back from that course going "that's it—I want to be an ecologist." And then a professor in my final year undergrad showed an interest in me and persuaded me that I could write a proposal to get Ph.D. funding. He was like a good friend as much as an advisor throughout the Ph.D., and very driven by wanting papers and success to make his research group bigger, so you wanted to do well for him. But no one ever really influenced or inspired me specifically for writing—I think that's very much something that's just a product of all the other steps in the research process.

I don't know how I learned to write science. It's definitely through the Ph.D.—I don't think I had any proper clue about it before that. I guess it's very much self-discovery and getting to know what will disappoint your advisor. My advisor would give me a lot of comments on my writing; he was not just one of these people who would say "mmm, that's no good. Rewrite it." He would give very constructive, detailed criticism and then I'd try to develop my writing style so that when I handed something in there wouldn't be much correction or criticism or commenting required on it. And I guess you read other people's work, and you have a research group around you where other people are at the same stage or maybe just a year down the line, and you can see how they're writing so much better than you, and you think "how can I get to that stage?" I find it very hard, thinking about it now, to put a finger on how I learned to write. How much of it was my advisor, how much of it was just learning, or all those other things around me, in my environment? But, I think they probably all came together in some small way to improve my writing.

I would like to have the opportunity to write with a little more freedom. It would be exciting to write in a style that you're not familiar with for an audience that you're not familiar with. It might be challenging to try and adapt and broaden your horizons a little bit. Being able to adapt to certain situations and convey your message to different audiences would be a really good thing to be able to do.

I'm not a writer. I'm just a scientist, a researcher. I always just see writing as part of all the things I do. Writing isn't the prime focus. I would never describe myself as a writer, but I guess I do a lot more writing than most people do in their day-to-day lives.

LIZZIE

Lizzie's office is light and colourful, and she is too. Her energy and passion for her topic are palpable, and she draws you into her experience of writing. She describes herself, after some deliberation, as an evolutionary conservation geneticist, and one of the challenges she has faced, as an emerging scientist and a scientific writer, is establishing and managing relationships—with colleagues, with students, with conservation officials, and with the amateur bird watchers who have watched "her" birds for decades. She wants, more than anything, for people to notice what her community is saying through her writing. When I interviewed her, she was sitting on the cusp between emerging scientist and established scientist—moving away from her previous advisors and establishing her own ground. She's transitioning into a new phase, beyond just writing up research, to commenting and contributing to her research community in a broader way.

It's Time to Stop Just Writing Up Research. It's Time to Start Commenting

Almost all the research that I do is collaborative—I'd say there is very, very little that I would do just on my own, and the nature of those collaborations really depends on the project. For example, I did my Ph.D. about six years ago, and that research now has two prongs to it: one is following up on some of the research I did in my Ph.D. in a collaborative role with a current Ph.D. student of my former advisor. And then also the project I did for my Ph.D. was very global and now I'm focusing it more locally. That research now is sort of mine and part of my research programme, so I no longer collaborate with my Ph.D. advisor, but I collaborate with other people here and internationally. I also do a fair bit of conservation genetics work, and I do that in collaboration with a government body. And I'm just wrapping up the post-doctoral collaboration with my post-doctoral advisor and now again launching that off into more of my own research programme. But that's through students and other collaborators.

So you could say I'm emerging into a new phase where I'm no longer the person working with somebody else's ideas and concepts and becoming one where I'm taking more of a leadership role. That's what I'm gunning for. It's not an ego thing, but if you want to establish your research programme, that is what it's about. I mean, in terms of authorship issues and things like that—at what point do you cut the cord with that previous advisor?

I'm also moving into the stage of my career where it's not me doing the work; it's the students doing the work, so I get to spend less time in the field. The vast majority of my work is actually done in the lab, but I haven't held a pipette in quite some time. Because I'm building on things I've already been working on, the samples are already in the lab. So a student can come in to do a project without ever seeing the species that they worked on. And I don't like that at all. There's a huge disconnect if you don't know your study species. I currently only have one master's student right now, but she's working with a recovery group for a critically endangered native bird, and so I've got her spending time with the conservation folks in the field and then also there are birds in captivity so she's spending time with the managers of those captive facilities as well. It's really important for students to get that interaction. Even though she doesn't need to go for her project, I think it's really important for her to know what she's eventually going to be writing about.

When you are a young scientist, you are usually species driven—you're interested in whales or you're interested in birds, and you generate your questions around your species. But as you grow up in science, you start generating your questions first and then looking for model systems in which to address those questions. I was the classic example of that. I was like "I like whales, all cool, there's these new genetic tools—I'm going to use these tools to answer these questions about whales." And I started thinking "well actually these are the questions I'm interested in. Seabirds are the great model species." So my doctoral project started with some ideas that my advisor had been thinking about and she was the one that pointed me in that direction, but then ultimately where the thesis went was generated by me. In my field that's usually how it goes. There will be some research proposal that says I want a Ph.D. student on this project, but then ultimately you're handing that over—the Ph.D. student's going to sort the details.

Let me walk you through my writing process. It's a long story. I work with a critically endangered bird, and one of the questions around them is that they occasionally hybridise with a self-introduced species. So part of the question with these guys is that, because they occasionally hybridise, there are people who think that they are not worth anything. That, from a conservation perspective, there is no reason why we should put any energy into these birds. And so for me the question was "Well are they or aren't they?" When I went into the project, I thought there would be some evidence of what we call introgression because what happens is, if you've got the two species and they mate and form a hybrid offspring, that's all well and good. But if that hybrid offspring then goes back and mates with one of the original species, that's how you get the DNA of one species into the other. So what I figured I would be doing is I would be talking about the conservation value of what we call a cryptic hybrid, which is a bird that looks like one species but has the other species' DNA. And I thought, okay this would be a really challenging project because it is not cut and dry. And in order to do that, we needed to develop some genetic markers to be able to correctly assess that because a little bit of work had been done previously but the sample sizes were low and the marker was inappropriate. But I insisted that we needed at least to see that data out, we didn't want to just jump blindly into a new type of data. And also I initiated a relationship with people in conservation management. To me, relationships are really, really important, especially in conservation. As a conservation geneticist you can publish a gazillion papers, but if a conservation manager wasn't part of the process or wasn't involved in the development of the conservation management recommendations, you might as well have not even done it.

I said "okay who do we talk to?" So then I went down and met with him, got to see the birds—really just had a conversation. And then shortly thereafter there were two students who came over on a summer exchange from overseas, so I took them down there. The same sort of thing: "you've got to go see these birds." And for me a lot of understanding whether this introgression is occurring is about the behaviour and about the management that has happened with these species, and all of that information is in the grey literature. It's in conservation management reports and things like that, and so the only way you're going to access that is through those people. There is a whole lot that isn't written down which you can read between the lines when you are in the know, but unless you are in the know, you have no idea.

For example, with a species that has a recovery plan, theoretically the recovery plans are meant to be published. They may or may not be accessible through the conservation management department. For example the '98 plan is available but the 2001 plan isn't, unless you go and talk to them. And every year the recovery group meets and there's an annual report. Those annual reports you could only access through the recovery groups. That's where the nuts and the bolts of everything are; the people writing the reports have this knowledge in their heads. And so in addition to going down and meeting with the local expert and then bringing the students down, at the next recovery group meeting I was invited as an observer, because I really wanted to see how this recovery group worked. Then I was invited to present, and now I'm occasionally brought down when they've got questions. So I'm communicating with some key people in conservation management, but I'm also communicating with people in the local community. And to me it's one of my proudest achievements, to be honest—and this paper that we've been writing is such a rich paper because I understand this system very, very well. And it's because of the relationships that I've built. I was a bit nervous about giving the studies species summary to the guy from conservation management, because he is very pedantic and very particular and will correct you if you say one word wrong about the history of this species. And he barely touched it. And I was like "Hooray!"

I mentioned what I thought would be happening, which was that I would end up debating the conservation value of these cryptic hybrids. Turns out these birds are genetically pure, as far as we can tell. What we thought was occurring isn't occurring. Which is why it's so interesting because it's like: How come? Because it is the "why" that's really meaty. And answering that why is what's required all this knowledge that I've purposefully gathered, but not really knowing why. And that then led to a collaboration with another colleague who's a statistician. I said "okay we need to analyse some of this data that's been kicking around in spreadsheets for twenty, thirty years." That's why I also like working with recovery groups, because if they're a good recovery group, they've got amazing databases. And it's all well and good to say I'm going to work on this species and go and collect data on them for a field season or two and try to infer something about the evolutionary history of that group. But when you've got three decades worth of data—you know? It took me a year to vet that data. I don't think I'd ever do it again, it would be a post-doc or a Ph.D. who would do that for me now. But at first it was just me. And I'm really, really proud of this particular piece of work. And the species in itself, it's a critically endangered bird, but nobody knows anything about it. And there's a lot of misconception about their genetic status. So I'm really, really keen to get some national coverage. I'll be really frustrated if it doesn't get picked up by the media.

Vetting the data is a huge deal because it's very subjective and that's when I knocked on the statistician's door and went "hey, what do you think?" And

Chapter 5

I said "if this is of interest to you then I would like to invite you to be on the paper. If you're not interested, that's fine." Sometimes co-authors can get ornery if a statistician comes in at the end and gets put on as an author. So I had to talk to each of the authors, and I was first author so it was ultimately my decision, but I said "you know, as far as I'm concerned he's an author. If he's going to do something for me that I can handle, then you know we could maybe debate it. But I cannot explain what he's done and not sound like a complete moron." And so that was something we had to talk about. I'm a really open, honest person, I don't do anything remotely sinister. I go "this is how it is" or "this is what I think" and everyone who is involved in this paper is the same way. So it was a really painless process.

So then I wrote up. I stayed home for a week in my pajamas and wrote this thing and just said this is like my Ph.D. all over again, this is crazy. And it was as dry as straw. It was so dry. I've talked about this research a lot, I've presented it at conferences, I've presented it at invited talks. And so I had, certainly, the abstracts and outlines from the talks that I had given. So I certainly had a very good idea of the structure of the paper, but mostly it was in my head. I said to a friend, "it's all in here. I just have to get it down." And that really is why it was a week, in my pajamas, writing. Because I just had to leave all this crap here and just write it. And it was—yeah, it was a painful process.

Almost always I'll start with figures and tables. So I was doing figures and tables, and then while you're doing figures and tables, you're drafting your methods and your results, and because we were going to a high impact journal I wanted really pretty figures. So I worked really closely with our graphics guy on figures. I spent a lot more time on that than I would for just an ordinary journal.

And then of course when you're writing up your results you're like "oh, okay, why did I do that? What was that about?" So that was all written over, probably like a month, in along with everything else I was doing. And then it was time to sit down and really get serious and write it, and so that was when I stayed home.

But I was a bit funny with writing this one because usually I've embraced the shitty first draft better (you've read *Bird by Bird*,[9] right? Everyone should read it) and gone "yep, yep, yep, okay I'm going to say something about this," and I'll literally type that: "I'm going to talk about this." And then write go "Now I'm going to talk about this." And that will help me with my structure. If I can't think of a word I need I just write **word** in the sentence and keep going and come back to it later. But with this one, because I knew that I really just had this week and if it didn't happen I was going to rip my hair out, I was a little more pedantic—more like "I'm going to fight with this sentence because I need to get it right." When I said it was painful, that's why: because I was so determined just to get the damn thing done. And then when I gave it to my colleague I said "you will fall asleep when you read this. It is so incredibly true and factual and bang on, but it is boring." And she came back and went "yeah, it is."

And she goes "It's all good, but you need all this, you need this information" and I said "I know, so I don't really know what to do with it." But I also wasn't worried about it because both my co-authors are flashy kind of writers, and I thought now I have other people who are going to help me flash this up. And that's certainly what happened. So then it went to the co-authors and they said "Oh yeah, put this over here" and, "ah, you need to spin this a little better, maybe just change this paragraph to change the emphasis so you're focusing on this aspect of it." And that was a really excellent process. And that was really a three-way between us to jazz it up a bit. And a lot of it too, I really enjoyed—I really had a lot of fun with that process because they would say "okay, try this" but I was the one who actually did it. So I don't feel like I wrote a boring paper and they made my paper sound good. I feel like I wrote the boring bits and then *we* made it sound good. It was a real team effort.

Sometimes you just need to get the damn thing out. I think I learned that during my Ph.D. I was doing this particular project, my advisor wasn't super helpful, and I looked at what we had and went "okay, there's enough data here for a paper. I think I should write up this data, and we'll submit it to this particular journal." My advisor actually discouraged me from doing this. She said "I don't think you have enough here, I think you should hold off," and I said "no, I'm going to do this." And it became my first first-authored work from my Ph.D., and everything that I have done since then (and also all her current students are doing) stems from that paper. Just get it out.

To go back to this recent paper, I was writing for politicians who make management decisions about money, I was writing for the conservation managers, I was writing for the faculty who often say "these birds are a waste of space and this is a really good example of what we call a hybrid swarm," and they're not. And I was writing for conservation biologists who were interested in hybridisation. There's a whole field that's interested in hybridisation and conservation and I was writing for that community. And then also there are spin-offs, which is just evolutionary biology in general, so there's a whole layer of people that that paper was written for. For example, there's a section in there that I wrote as a response to communications that I've had with academics in a particular country. And I wouldn't be surprised if the journal tells me to take it out. And a couple of co-authors said, "Do you want this in here?" I said, "Well I'm going to try. If they take it out, I'll write it somewhere else, but let's just give it a go."

The politicians won't read the journal but I'm hopeful the media will pick it up. The conservation managers will read it because the academics they work with will read it. And the evolutionary biologists and conservation geneticists and those hybridisation types will all read this journal. Evolution biology is a massive field, but that little subsection that are interested in this sort of stuff, they'd already be reading that journal anyway. It might get completely lost in another journal. And we want people to read this one. It's not just to put on my bloody CV, this is one I want to be read. So I was really going for the broadest possible audience.

I'm doing an assignment with some juniors right now and there's a paper called "How to Read a Paper" and—have you ever come across it? It describes this first pass, second pass, third pass approach, and we've started introducing it at second year here. We have built it into either lab exercises in one of our sort of lab-y kind of courses, and into a lab report exercise on the ecology side of things. Every student we have in the department ends up in one of those two courses, and we wanted to capture as many students as we could, really embracing that idea that you have to teach writing within your field. So what I'm now doing in third year with students who got introduced to that idea last year is we're building on it, and I'm saying that in order to figure out if you understand a paper you need to be able explain it to someone. And so I had them going through the first pass and the second pass in a tutorial yesterday. And part of the first pass is "is it well written?" You see, I think they're your audience too. It's not just your post-grads and the academics, it's also the undergrads.

One of the things that we said to our students is that writing is a skill. It is a skill you hone and you develop and the only way to get better is to practice it. And I used that book (*Bird by Bird*) as an example and said you've got to start somewhere. And I was really, really lucky in that when I was writing up my master's I had a mentor, a very, very good writer, and he held my hand through that process and I learned a lot about good writing from him. And I got really, really lucky during my Ph.D., again. My advisor wasn't very helpful—she would correct things but not tell me why—but I had a close collaborator and he would tell me why. He would edit and he would tell me why. And my writing improved so much during my Ph.D. because of him.

I got the kick in the pants that I needed during my master's, but I was still really slow and really pedantic. It would take three days to write an abstract and people would be like "oh yeah, you know, a few more years from now and it'll take you 20 minutes." And I'm like "no, that cannot possibly be true!" And now I'm popping them off. And I think communicating clearly is really important to me and I'm also a pedantic kind of person and quite literal. So writing well is important to me. It is a skill I've purposefully worked hard on. And I think, also, when I was at elementary school it was during a phase in the 70s when they didn't think they had to teach grammar, you know, and all that nonsense. So actually I can write you a nice looking sentence but I can't tell you what the noun is, what the adjective is, I don't know. And I remember taking an English class at university and I was lost. I was like, "I don't know what any of this stuff is." I just did not come from a really strong place, and it's literally just been practice and good advice. And lots and lots of support.

Now I think I'm pretty good. I don't mean it in an egotistical kind of way, I think it's just this a skill that I really value having. And I think also it's not just that I want my students to publish, I want my students to be good writers. Because I don't care if they become academics, or managers, or moms, I don't really care what they do, but I want them to be able to communicate, you know? And it doesn't really matter what you're communicating; if you're a good writer and you know how to tell a good story, it doesn't matter what the story is.

I'd probably say the intro is the hardest thing for me to write—not so much the discussion. Because there's so much and you've gotta cut through all the crap. If it's something you've been working on for a while, you've spent a lot of time thinking about it, you're really aware of the literature. But then when you go down to write the paper—especially if you don't have a good idea of what journal it's going to, how you start it can be so key.

I think it's framing the paper . . . I mean it's so obvious. The intro is "this is why we need to do this, this is what we're going to do." You know it's really, really basic, but actually snipping out the extraneous stuff and getting it down into "well, what is the relevant background information—what is it that I am actually trying to do here?" can be a struggle. All this other stuff is interesting, but what am I trying to say? And for my big paper it was about capturing the audience right away, because I want to make the biggest possible impact in these first two sentences. And actually the first two sentences weren't too bad.

The easiest kind of writing is writing out a research proposal. It's so easy. "I'm going to do all this great stuff!" Coming up with the idea is easy and is definitely a lot easier than writing what you found!

The relationship between writing and science? That's a pretty open question, isn't it? It's co-dependent really. If you don't write it, it didn't happen. And if you don't communicate it well it doesn't get passed along the chain. And from a teaching perspective, that's what we're telling our students more and more. You have to embrace this skill because you need to be able to communicate what you're thinking or what other people have found or why you're doing what you're doing. Even with this example of the work I have been talking about, I need to get that paper out so the recovery group can go "it's *this* paper, it's not just what Lizzie said."

So the status of what I'm saying changes. I'm going to be really pissed if this paper doesn't get accepted, because I'm really fascinated to see how it goes. I'm

Chapter 5

curious about how attitudes might change when it's not just me saying this, it's my colleagues and it's been peer reviewed.

I had a paper come out in *Biology Letters* last year and it got pretty good media pickup. It was about the rediscovery of this bird, merging ancient and modern DNA. And again that was written with a very specific audience in mind, very punchy. It went to *Biology Letters* because I needed it for my research rating; I needed to be able to say "there's my letters, here's my international coverage, here's my national coverage. I've got some experience with media interviews, tick, tick, tick. Done." M, my collaborator, also worked on that paper. She doesn't work on birds, she's like "oh, no-one's going to care about this" and I said "It's birds. People love birds!" And they did; the media here really responded. I'd presented very similar data at a conference in Barcelona, and no one really cared. And that's what I think I found so interesting, that among my peers, they're like "oh yeah, that makes sense, that this presumably extinct thing and this living thing are one in the same. Well, yeah, okay." I did find that kind of interesting, that my peers didn't give a shit.

But this paper I've been working on should be interesting to both communities. It's interesting to the conservation community because it's a really nice, good news story, but to the academic community it's a, "well how the heck did that happen?" kind of story. Like a puzzle, a mystery. And we gave this paper a very jazzy title—that was very intentional as well. We had a lot of discussion around this. Apparently, with those high-end journals, if there's a species name mentioned in the title, they won't even look at it. That's what I've been told by the people that publish in these journals, and that's good to know. Whereas a related one that went to *Conservation Genetics*, I was loud and proud about writing the species name in there.

I'd say that I was definitely learning how to write science during my master's and Ph.D. I already sort of knew how to write, but in my master's and my Ph.D. I was learning how to write science. I think particularly because I was writing papers. Now in science it is very rare that anyone will actually write a standard thesis; usually it is two to five data chapters sandwiched in between a general intro and a general discussion. And so I think the scientific writing skills that we're getting now at graduate level are probably different to what scientists would have gotten in the past. I think I'm honing the skills, learning how jazz it up. It's about learning those little things, the tricks of the trade. You've got this rejection letter from one journal, how do you turn it around to be something positive? Like with the recent submission, the one I've been sweating over, in my letter I incorporated the fact that I got feedback from the senior editor of another journal. I wouldn't have had that savvy during my Ph.D. I remember I did some stupid things, like I submitted an article to a journal and didn't even write a cover letter for it. Now I wouldn't even think of doing that.

So I'd say that this next stage is just getting savvy and honing your skills and picking your battles and being really strategic. That's not just the way I approach my research, but the way I approach my early academic career. It's all about strategy. And it's all about not just ticking boxes but using your time and your energy efficiently, so you can have a life outside of academia.

I'm getting to the end of that stage now. Right now with M and another friend of mine (who I've yet to publish with, but we've always known that we would eventually), we're ready to write a reply to an article that is absolutely pissing us off. It's time. And I know it. I go "yup, it's time to stop just writing up research. It's time to start commenting." I guess you just feel you've got the expertise to see something and go "that's just wrong." With this one particular paper, all of us are reviewing articles where people are citing this guy and it's just a pile of horseshit really. It's pissing me off. I feel bad that I'm reviewing this Ph.D. student's paper, and they don't know any better, so I'm just this nasty reviewer that goes "well I know you cited so and so, but so and so's on crack." Well, of course I don't say it like that—it's much more professional, trust me. But that's what I'm thinking. I said, "look you guys, we need to write something because I want to give those students something else to cite." Hopefully someday someone will invite me to do a review, though I've heard it's horrible so I don't think I really want to do it.

I think you become savvier about not just being a good writer but at writing to an audience, that you're picking your audience and you're writing to that audience. And I think that's where I'm at. I know who these people are who will keep writing the same papers to the day they die, the same sort of formulaic kind of stuff. I want to get savvy and become more proactive than reactive.

LEMROL DARNEL-GAN

Lemrol is a senior academic, past retirement age, but still working full time and taking an active research leadership role. It is hard to pin down his discipline: his overarching interest (and the discipline in which he began his career) is applied physiology, though his current primary interests are animal welfare science and bioethics. He speaks with authority, in long paragraphs and with precision. He somehow finds time for me to interview him three times (for a total of three hours), because there is so much to discuss. I chose this small part of his narrative as the final narrative in this section on the development of the scientific writer because, in a sense, he picks up where Lizzie left off. She is just realising that she has to think bigger, start engaging with and challenging the central

Chapter 5

questions of her field. Lemrol's end-of-career narrative shows what the career of someone who's done that, who's looked broadly and creatively at their discipline, can become.

Everything I Write, I Still Learn Something New

By the luck of the draw, by the accidents of circumstance, I've done a lot of writing. I did my final year at high school twice. I went back to repeat the year, not because I had failed, but because I didn't do particularly well. I had great learning problems at school. I'm mildly dyslexic, and in those days, the 1940s and 50s, they thought you were intellectually handicapped. It also gave me an opportunity to do the same subjects again, but I elected not to do English, which was really stupid, although I didn't know it at the time. At university I started an agricultural degree, changed to science at the end of second year, and majored in physiology. So I had a BSc, and that was followed by a final full year of research at honours level. That was when I had my first introduction to the contingencies and the difficulties of writing about scientific subjects. I was extremely lucky, in that it was a small, fully residential university and so you virtually knew everyone. And I had very nice friends, some of whom were in the humanities. One in particular, very kindly, when I was writing my honours thesis, took me under her wing, and helped me with English expression, because I was struggling with that.

What I found difficult was the focus, the precision, the use of language, how to express what I was trying to express. I mean, you thought you had it, and you'd write it down, and someone would look at it and say "I don't know what you mean." The way I say it to my students now is that "I was not born writing the way I write today, I had to struggle and learn exactly what you have to learn." And this is where I usually quote T.S. Elliot's poem "Four Quartets: East Coker" to them:

> To arrive where you are, to get from where you are not
> You must go by a way wherein there is no ecstasy."

I say to them, when they're writing theses and assignments, that there are two processes going on, unless they're exceptionally lucky and they have done humanities and actually have a facility with English already (and even then, they still have to learn the precision of scientific writing). They are working out what their ideas are, and, at the same time, they're learning the careful use of English so that they don't overstate or understate the situation; also, so readers who don't know really what their ideas are can understand them from the very beginning to the very end of what they have written. Those two processes are so overlaid that you can't really separate them.

It's a long and painful process. And you have to learn from ground zero, which is what I had to do too. You must go by a way in which you don't know. This is why you have to trust that your advisors know what they're doing on your behalf, because you can't get someone to understand what they will understand at the end of the process when they have not yet gained that understanding. And I still learn. Everything I write, I still learn something new. There is always another way of putting an idea, or a fresher way.

In the early stages as well, you do not have a vision of the subject. You've been gobbling up all these references, which you're reading to get facts. Then you come to know the facts, and you become burdened with facts. And those advisors who are not that experienced in scientific writing and in scientific research say to their students, "you've got to write your literature review first. You've got to do your historical review of the literature." And I tell these students that it should be called an "hysterical review" of the literature, because, if you write a historical review of the literature reporting only facts, you will bore your examiners to the edge of insanity before they even get to read about the research you've done. So, you do all the reading, but the last thing you write is the introduction to a thesis. That is unless we can get them involved in developing their wisdom of the subject early, developing a vision of the subject that's publishable. If the introduction is not publishable, we don't think our students have done as well as they might. I've had several Ph.D. students who have co-written three to six review articles with me as part of their introduction.

I was extremely fortunate that I lucked into a productive area for my Ph.D. My first advisor was unwell and he died 18 months after I arrived. So, I ranged around to find something that I could do. I was fortunate to find a very productive and interesting area which included human beings. By the time I finished my Ph.D. and landed my first job, I was much better at writing. But oh boy! I was still raw—my style was very verbose and repetitive. There was a guy there who was head of the biochemistry department who was brilliant at science but often inept with people. I learned a lot from him! He was a more senior staff member than I was, and I was a pretty young head of department, so the director of the institute, who was not a physiologist, said I had to show my papers to this biochemist before sending them to scientific journals. And he would take a sentence that was maybe 25 words long and he'd reduce it to 10. And it would be much clearer. But as he wasn't a physiologist either, he didn't understand how it should be written for physiology journals, creating other challenges for me. But that's another story.

So I'd written my thesis, and six papers that came from that work. Everything you did, you had to publish. I mean, you *have* to, because this shows its value. If you can't publish it, it hasn't been worthwhile doing. So everything

Chapter 5

you'd done should lead to a paper. Being inexperienced, you'd just write it the best way you could. It would then go out to referees, and the referees would come back and say "well I don't understand this," "it's not clear what you mean here" and so on. You can learn a huge amount from referees, even if they misunderstand. Because, if they have not understood what you've said, and they come back and say "oh, this is rubbish," you have to ask yourself "why?" So, every time you get a paper back from a referee, it's a learning experience—as long as you're not someone who reaches for the irate button.

Sometimes reviewers talk about writing style. Editors occasionally do as well. I published quite a few articles in one particular journal, and there was a period there where I'd get every manuscript back after it had been accepted for publication with red writing all over it. About 30% of the comments were helpful in making the writing clearer. But 70% didn't make it any clearer; all it was doing was putting my writing into the editor's style. He clearly didn't have enough work to do, despite being a full-time scientist as well! So eventually, after about eight or nine papers where he had done this, I wrote to him and said "I know you're trying to be helpful; I really appreciate the effort you're putting in; but to be perfectly honest, I think you're going over the top, because I believe you are now trying to convert my writing into your style. I'm very happy to accept the things that really do make it clearer, but I frankly want to retain my style." He got back and he said "yeah, yeah, fine. No problems. Take or leave what I say as you see fit." This was about five years after the Ph.D.—I was confident about my style by then.

But if you'd said "what is your style?" I wouldn't have been able to tell you. It was still evolving. If I looked back at some of my early papers, I would say "oh, I could say that much better now." And, so, you keep on learning, as you get more involved, more experienced, and as the breadth of your vision of the subject widens or deepens. Then you have the confidence of that experience behind you, and you are actually putting your writing in that context. In the early stages, the writing is part of working out the ideas. That hasn't been the case with me for many years now.

Once you get enough into the subject, you get to the stage where you have thought a lot about it, so when you're sorting out the results you are also looking at the interpretation. And so at an earlier stage you'd get all the results, and then you'd say "oh god, what the hell does this mean?" But then you get to the stage where you've worked out what it means before you start writing. You've discussed it with your colleagues, you're saying "oh this fits in here, and what that's showing is this," and so on, so that, in a sense, you have a much earlier idea of where you're heading. But it sometimes happens that in the process of writing, and then pulling in the references to give the embellishments and the support, or the caveats, you suddenly have a fresher idea than you've had to that point. The writing might take a different direction. So once you know your field well, you often work from a base of ideas that you put together at the very beginning.

I think a way of seeing it is to look at the different sorts of articles. When you first start, you're writing original scientific articles, where you've done the research, you've evaluated the results, you've looked at them in context, and you've written the paper which explains the significance of the results. At this very early stage, you're learning how to use English, and you are working out your ideas as well. So you learn something new about English expression every time you write. You learn less and less about English expression as you get more and more experience, but you still learn something every time.

And, as you get further on, having written many original scientific articles—and this may be peculiar to me—you start asking "how does this fit into a review article I could write, to carry the subject forward?," rather than, "this is where we've got, and it stops here." You think "What do we know and how does that take us forward?" I advise junior faculty to think "what review am I going to write?" because that helps them crystallise their vision. Now, you have to be very careful that your vision is open, that it isn't closed and you're not disregarding anything that disagrees with what you want to put forward. So yes, you develop that vision. And that vision is built on original research and, we hope, is also pointing to fresh ways of thinking about the subject.

So, reviews at their best provide a solid foundation together with fresh conceptual frameworks to help the subject move forward. As you get more and more experience with writing papers, you get invited to present longer articles where you actually begin to tell a story. And that tends to involve drawing information together, and putting it out as important ideas for people to have in their minds to stimulate other research. So, I started writing reviews, or at least thinking about writing reviews, many years ago and now review-writing is my major activity.

After 28 years as a head of department and researcher, I became a full-time scholar and academic. This was 11 years ago, and it has just been absolutely wonderful. What I've noticed, and this is something I've heard other people comment on, is that I began to say: "well, part of my obligation now, as someone who has the time to reflect on the wider background to the subject, is to draw on my professional lifetime of experience, to see how that can be brought forward into the present, to give a basis for going forward; not as a strait jacket which rigidly holds onto particular ideas, but a launching pad for originality." And that's where it's been real fun. I've really enjoyed writing reviews and books, and co-editing other publications for international organisations.

Physiology, to me, is just fascinating. And what I really like to do is to get across the fascination of it. And it's not just fascinating. It's really useful.

Chapter 5

The ideas that engage the imagination and lead to fresh directions in research are the ones that fascinate. I'm saying "have you thought of this this way?" For example, take the question of fetal consciousness. I presented the evidence for it, some of which is quite controversial, at a scientific conference and someone said "but isn't this relatively well known?" And I make no pretence about it: "Yes. It's been in the literature for many years. Fetal physiologists have known this for 25 to 30 years." But one of my colleagues said "yes, this is very well known, except for this new synthesis. What Lemrol has done is completely fresh; no one had ever drawn it together that way. And he's the first one who's dared to say it." And it's led to all sorts of things, like special global guidelines for the slaughter of pregnant animals, to make sure their fetuses are managed in such a way that they can't suffer. This kind of writing can lead to all sorts of fascinating new directions.

So, at the beginning, when you start writing science, first of all, you're learning to write. Then you go through a phase where you are writing to discern what your ideas are as well as to understand your field and learn to write about science more effectively. And from there, you go on to writing reviews, where, in a way, you know what the ideas are at the outset. You know the field so well that you write and your ideas are already formed. It's simply a matter of putting them out there. And then finally you get to a stage—or maybe not finally, maybe there is another stage after that—where you are providing a platform for something that has huge implications in lots of different ways.

I learn something new every time I write. That is what the enjoyment is. For example, I have another review that I have in mind, where I'm not absolutely clear what the outcome will be. Well, I am clear what the outcome is going to be, but I'm unclear as to how I'm going to express it, if you see what I mean. The question is "do you need to be conscious in order to learn something?" People say "well the fetus must be conscious because it recognises its mother's voice after birth." Well actually there's no physiological reason why it needs to be conscious for that to occur. But I need to explain it physiologically. And that's what I've set myself the task to do.

But there's another project, this time on animal welfare, which has a rather different focus. This reminds me, that maybe we need now to think about the breadth of subjects, how my interest in broad issues arose, and the impact of that on even wider issues. Much of this has come from other activities that I've been involved in.

So fetal and neo-natal physiology were my major areas when I worked in Scotland, when the context was the causes and prevention of lamb mortality. Also, the fetal/neo-natal area gave me a link between agricultural and veterinary sciences, science, and medicine. Now, part of my neo-natal work led me to look into stress physiology and pain. And so, when I came to my current university,

I had also published in the area of the impact of routine husbandry practices on animals, looking at behaviour, pain and stress responses. As a busy head of department, I couldn't manage fetal and neo-natal physiology projects, because they're pretty much full time, but I could do focused projects with masterate students. These were on different aspects of assessing how much pain was caused by these husbandry procedures, and finding methods that were practically usable and not costly on farms for relieving that pain. So that got me into the animal welfare area.

Now this is where opportunity can lead you to different writing styles, different writing areas. So I'm just telling you the history here. If you'd said to me 21 years ago "this is going to happen," I wouldn't have believed you. Once here, I became involved on behalf of the Royal Society in establishing a council for the care of animals in research and teaching. I was its executive vice-chairman for many years. So that put me into the area of bioethics applied to animals used for science and for other things. One of the ideas I really pushed was that scientists should become ethically literate. Previously I started a bio-ethics discussion group and afterwards I had a lot of work setting up conferences and so on. Now, that involved a completely different sort of writing. Getting ethical ideas across. My inaugural address at the university was a different sort of writing as well, as it was the first time I had nailed my colours to the mast, where I was saying that scientists have to embrace the ethical dimensions of science.

This path I've taken is, perhaps, unusual. But it is usual in the sense that there are always a few unusual individuals who have such broad interests. This doesn't make me wonderful or special. It just means I have a different orientation to knowledge and thinking that takes a wide compass. Other people can't cope with that. Now that's not a negative thing. Other people prefer, or are more satisfied with, a more precise focus. It may be because they are a bit nervous about stepping outside their comfort zone, or it may be because they just really like the pursuit of that very detailed knowledge. And I'm not judging them one way or another. Both approaches have a legitimate place in science.

My interests have always been extremely wide, and include religion, culture, history, economics, philosophy and politics. So, as chairman of an animal welfare advisory committee, I found myself writing for lay people. The areas of writing sometimes really surprised me. At one stage we had a challenge from one of the industry boards when we were preparing a code of welfare which was a major part of the remit—and still is—of that committee. They sent in this challenge from one of the leading lawyers in the country, about the fact that we had interpreted the Animal Welfare Act the wrong way. And I said "well, I don't think we have, actually." But I couldn't say why. I found myself writing a 20-page analysis of the Act demonstrating that we had done exactly what the Act intended in its spirit, and in its detail. So I found myself then writing on legal matters. That was fascinating. Incidentally, no response came back from the lawyer!

My late first wife, who was from India, wrote a book which was an analysis of crossing cultural and religious barriers. It was an analysis of why it was that something we found so easy, everyone else seemed to find so difficult. We didn't see it as a barrier, really. But you don't realise how difficult that can be until you look back at the journey you've actually travelled yourselves. It was a delightful thing to do. It has been immensely helpful in all the other things I've done, and is very important in my writing. What I learned through her writing of that book was the importance of finding ways of really truly putting yourself in the position of someone who doesn't have your primary assumptions. And the point there is that you start from a position of respect. You don't start from a position of "I'm better than you and everything that you think makes me feel good, because I'm right and you're wrong." And so that has given me a facility of actually taking what people have written, or what they say, and finding out where they're coming from.

As an example, when we were looking at religious aspects of slaughter, we sent out a discussion paper asking for input on particular aspects of religious slaughter, before the national organisation came to a view on how religious slaughter should be managed from a welfare point of view. We had a significant number of submissions from all sorts of groups, including the Jewish community. So I analysed all of the submissions, not just the submissions from the Jewish community, and presented them from the point of view of the people who had made those submissions. Again, this is the business of writing. After I had written the part for the Jewish community, I phoned one of the rabbis who we'd been in discussion with, and who had made a submission, and I said "look, this is what I've prepared on your behalf to present as a statement of your position." And I read it to him. And he said "are you Jewish?," which I took to be a very great compliment. And I said "no, but I've spent virtually all my adult life trying to put myself in a position of other people who have other views from my own." That undoubtedly has influenced my writing style.

If you're only interested in the science, that is all you will do. And you will feel really uncomfortable stepping outside what you know. But as I said, I'm an integrative holist person with eclectic interests. And so, it wasn't that hard for me to do, because even at school I had a personal interest in studying world religions. And then, of course, these interests grew stronger when I met my late first wife, as she opened up a whole world of completely different orientations to spiritual expression and human existence. So what I am able to say is, I'm an ethically literate scientist who feels completely comfortable operating in the humanities and other wider areas; not claiming to be an expert in them, but completely comfortable to operate in them, and draw out of people what it is that interests them.

CHAPTER 6
THE POETS

> The equation is poetic.
> — Catalizador

> At the core of science, science is an art. And, you know, one of the hallmarks of art is beauty of expression. Good science writing should be completely unambiguous, but it should also be beautiful.
> — Senior Scientist, Nutrition and Physiology

Much has been made of the division between science and the arts. Laura Martin (2012) goes so far as to suggest, albeit implicitly, that writers and scientists are not the same people:

> Good writers and good scientists share many attributes. Both care about their representations of the natural world. Both work constantly to improve their craft. Both care about clarity and about audience.

C.P. Snow, a writer of fiction and science, writing half a century ago, deplored this "gulf of mutual incomprehension" between what he describes as the "two cultures" as damaging to society as a whole. Lawrence Krauss, writing in *Scientific American* in 2009, suggests that Snow's call to experts in both fields, to build bridges between these two cultures, has gone unheeded, that the gulf between the experts in these differing fields remains firmly in place.

The senior scientists in this study, however, do not entirely support this proposition that the arts and sciences are starkly separated. Many of the participants showed a spontaneous interest in art, music or literature. Richard's comment (Chapter 2) seemed to speak for many of the participants in this study:

> I love language, and love to read poetry and things like that. And most of my scientific colleagues who are leading scientists throughout the world are like that. Talented people are interested in stuff. How could you not be?

The majority (76%) of the senior scientists I interviewed described themselves, in some way, as regular readers of fiction. Many of the senior scientists I interviewed moved away from the questions about science writing at some stage in the interview into an enthusiastic discussion of a book they'd recently read, or a book they had long loved. Several of my interviewees' offices were strewn not just with scientific journals but with literature.

Laura Martin (2012), in her discussion of the perceived separation between science and the arts, attributes the responsibility for this division to our education systems (see Chapter 7), but the results of this study bring such a statement into question. One issue that arose in the interviews was the whole question of creativity and science. Clearly, a full discussion of the relationship between science and creativity is beyond the scope of this study, but scientists' perceptions of this relationship are not. For several participants in this study, the sheer open-endedness of creative writing at school was both puzzling and, in a way, frightening. Not until they were able to find something concrete on which to pin their imagination were they able to engage creatively:

> I never really enjoyed [writing at school]. Never. Until later. And I guess that's partially because it was creative writing, and that's quite different to the sort of writing that I do now, which I enjoy . . . I can certainly see a creative component in [scientific writing], but it's still centred around something tangible; I don't have to make it up. There's still data or theories that I can use to develop my story. (Senior Scientist, Asthma Research)

Within this context, many scientists saw themselves as engaged in a creative endeavour, which involved creativity in writing. Even Mason (Chapter 3), who was generally disparaging about the arts, saw writing papers as creative. In response to a question on the survey, "Have you engaged in writing a piece of creative writing or creative nonfiction in the last 6 months" he wrote: " I object to this question. Obviously scientific papers are creative non-fiction."

The notion of creatively constructing a story or a journey narrative was central to many of the interviews:

> A good [science] book takes the reader by the hand, and you start a journey together—even if it is a journey to formulas or topics, it is a journey. And you have to organize this journey, and this journey can be pleasant or [not], and that depends on the writing. I think this is part of the writing process, creating a story. . . . So I like writing—it's part of the creative process. (Senior Scientist, Biological Physics)

One participant, after the interview, sent me an extended discussion of this idea of story:

> A good short story is circular. The protagonist starts at point A, is confronted by a situation that raises the question of how will they resolve it, and has an outcome, B, that takes the character to a new state of being, C, which somehow reflects on the starting position or has echoes of the opening. The same may be said of scientific writing, starting with X plus Y, the outcome Z is achieved, which is compared with where we started, and what the future holds. In both situations, we (the readers) are taken on a journey—from A to C. At point C, we can see now see point A behind us but in a new context, and, excitingly, can see C and perhaps other points (D and E) marching off into the distance (implications). A good paper/article/grant application or piece of fiction, takes us beyond ourselves and opens new thought horizons. (Senior Scientist, Neurobiology)

Which is close to Randy Olson's comment in his provocative book *Don't Be Such a Scientist*:

> Science and film-making . . . are both exercises in storytelling. And thus they conform to very similar rules when it comes to doing them right.

On a more frivolous note, scientists told me about their enjoyment of creating amusing titles for their papers, or introducing slightly offensive acronyms into their writing, and dreams of introducing and referencing "a completely spurious concept invented by a completely spurious scientist" (Senior Scientist, Chemistry). For some, the creativity of writing was a central part of the fun of writing—I'm thinking of James in Chapter 2, for example, enjoying the challenge of working out what information readers need at different points in a piece of writing. Some saw a need for more scientists to engage in this kind of creativity in their scientific writing:

> I think scientists would love to write a lot more if they could be trained to be creative in the way they write, and to respect it as a creative process rather than a chore. (Senior Scientist, Nutrition and Physiology)

Yet the issue of creativity in scientific writing is complex, and needs to be managed carefully in the context of gatekeepers' expectations of genre. As Burton and Morgan observe, less creative approaches to writing (one might argue, more formulaic writing) are associated with the authors' needs for safety, suggesting that those most likely to feel safe with a more creative approach are likely to be those whose reputations are already established. This was certainly borne out in this study: senior scientists were far more likely to take risks with style, and to do so intentionally. Yet even those who stood at the pinnacle of their field, such as Catalizador in this chapter, trod warily in the face of gatekeepers (reviewers and editors).

Beyond the issue of scientific writing and creativity, the notion of a separation between science and the arts was disputed in the activities of the scientists in this study. A small number of the senior scientists had studied literature at tertiary level beyond core course requirements, or had considered majoring in the arts. While the majority of the senior scientists did not engage in creative writing, a significant minority (17%) had been engaged with creative writing in the last six months, either on a private level (the "secret writing lives" as Poe et al., 2010 describe these activities) or in the public sphere, while several others told me about creative writing or creative endeavours (such as advising a writer of fiction or an artist on some aspect of science) in which they had previously engaged. As an example of private writing, one of the interviewees had written, for family and friends, parodies of Harlan Ellison, Wittgenstein, Plato, Thomas Aquinas and Ayn Rand. Several of the senior scientists mentioned an involvement with music, including writing lyrics, while others had published poetry or science-related creative non-fiction. One had a secondary career as a writer of young adult fiction.

Interestingly, while almost the same percentage of emerging scientists as senior scientists described themselves as readers of fiction, somewhat fewer emerging scientists I interviewed (only 11%) were involved in creative endeavours, either privately or publically. One of the emerging scientists had recently finished his B.A. majoring in English, but only two were published creative writers. We may speculate about the reasons for this: it may be that the pressures of acquiring tenure, for example, work against anything but a focus on professional reading and writing; however, emerging scientists working in a context which had no tenure requirement were no more likely to be readers or creative writers than those working within a tenure-based system. Such speculation is also undermined by the fact that 37% of the doctoral students interviewed—surely the group we might expect to be most focused on establishing their scientific credentials—engaged in creative writing of some sort. However, the size of the doctoral student sample was small, and so these results bear further examination with a broader sample.

For most of the creative writers in this sample, their creative endeavours were, in some sense, separated from their work as scientists. Even when they were working with poets or artists as science advisors or writing creatively about science, they saw this activity as distinct from, and not affected by or an influence on, their professional careers. In their discussions of these activities there was little crossover between the two activities; writing creatively did not influence or cause them to reflect on their writing of science, and their professional identity as scientific writers did not lead them to reflect on the nature of creative writing or art. I include in this chapter one short extract from a physicist who exemplifies this separation. While he has worked with poets on collaborative projects, and publishes his own poetry, his primary focus is observing a fellow poet come to understand something of the language of science and mathematics rather than engaging conceptually in this process himself. About his own poetry he says almost nothing. The second narrative, by a young food technologist who writes game books as well as poetry, attempts some analysis of how, for him, the different kinds of writing have affected one another—but he too sees the two activities as quite distinct.

However, two of the creative writers in this study did engage in integrated discussion about their writing across science and the arts. One was Elizabeth from Chapter 4, in her extensive discussion of directing her writing to a range of audiences. The other was Catalizador, a poet/playwright and a research chemist, who discussed various attempts to bridge the gap between his science and his work as a poet. His narrative, which was quite different to any of the others in the entire sample, seems a fitting final narrative in a book about scientists as writers, in the way it disputes this distinction between writers and scientists.

MICHAEL

Michael is a senior physicist who also has a national reputation as a poet. He has also spent considerable time working with artists and poets, as a science advisor on national creative projects, and he's a broad and perceptive reader. His comment that scientific writing is a craft, but not one that "leaves a personal signature," suggests a certain diffidence about his writing. And while he expresses uncertainties about engaging with new audiences, he is someone who is prepared to sit in the intersection between creative and scientific writing.

WHY DO I DO IT? WELL, I DO IT BECAUSE IT'S FUN, AND BECAUSE IT'S INTERESTING, AND I LIKE IT

I don't think I'm too bad a writer. There's no point in having a great idea if that great idea never leaves the four walls of your office, so it's got to get out there. And I think good ideas spread quite quickly in the scientific world. Getting them out into a bigger audience usually seems to be more difficult. And that's a barrier that you're always going to have, because I think as scientists you're trained to think and to communicate in certain ways, and it's certainly not the way that the general public thinks or communicates. And so bridging that is a difficult game—but not impossible. But it does make people uneasy, and there are certainly problems when there are political ramifications. If you think about what's going on in the whole climate change argument at the moment—I mean, scientists are trained to not be certain, scientists are trained to be doubters. And so, when a scientist says "we're pretty sure this is what's going on" then there's always something, there's always someone that's going to ferret around and find the 5% around the edges that we're not sure about. Which is what we're trained to do. But is it likely that that 5% of effect is going to be more important than the other 90–95% of the weight of evidence?

One of the things I've thought about is whether I should try to write some more "popular science" type articles, which I sometimes think would be a breeze—you know, write something that is aimed at a *New Scientist* audience. I've never done it—it's just a matter of time. But I think that would be an interesting challenge, as I'm not sure what the audience is there. My reading of something like *New Scientist* is that I'll read in-depth something that's close to my field, and I'll read every word about it. You know, if it's something about, say, cosmology (that's stuff I find fascinating), I'll read it all, but if it's something about evolutionary biology, which is interesting, I'll just skim it—I'll read the first few paragraphs and introduction, and look at the pictures, and see if there's something there—but I won't read it in-depth. And so I think it's a very difficult thing to try and write something that will sum up the field if you like, or talk about some interesting new results, in a way that will both engage the physical scientists that are out there, but would also be able to say something to the general readership of *New Scientist*—something that will hold their attention for a whole 500 words or whatever, I think that would be terribly difficult.

I try to write for pleasure. I read fiction all the time. And I write poetry. In 2005 I was involved with the "Are Angels Okay" project with the Royal Society, and I had a poem published in that collection. And I still just write . . . dabble from time to time. "Angels" has been a really nice project, and things that have gone on from it have been really nice. But I don't think there's any influence between the different types of writing I do. Certainly I can't see my journal articles influencing anything else I do.

Following up from the "Angels" project, I was advisor on science to a poet. So we've done at least three presentations together since that year—it's been a while since the last one. He's one of those guys who you just pick up with immediately. He thinks really hard about what he does and it looks so absolutely effortless. He's really thinking about new ways of presenting poetry, which I think is really exciting.

Being a science advisor to him was just talking about science. The "Angels" project was about getting 10 writers to write about, or come at, physics (because it was the international "Year of Physics") from their perspective. So he wanted to write some poems and we talked about themes, and it turned out that he actually liked physics and liked ideas, but always had trouble with mathematics. Mathematics is really the language of science and physics, and so getting to grips with that was actually a real stumbling block. In many ways, it's not just for this poet, it's for all our students. He wanted to look at a lot of the equations that we use and reinterpret them or translate/interpret them. And so that was the project we came up with and he wrote these 10 poems or series of poems about various equations. We would look at them and he would ask me a lot of questions about what they meant and what sort of things you did with them—why people thought that certain equations were particularly important or beautiful or whatever. That's a word that gets bandied around a bit, but there *is* a certain beauty in some of these equations. And it was a marvellous job, actually, it was a great project. You can contrast what he did with another poet in this country, who also wrote a suite of poems about science that were very, very different in character. But both those sets of poems are wonderful.

One of the poetry series included a beautiful poem about spiders' silk, about how a spider makes silk. The poet who wrote that series talked with a range of scientists about materials and things—and looking at them poetically. So a lot of his poems were perhaps a bit more prosaic in that sense, talking about everyday materials or coming at new materials. Some of it, I think, was a process of discovery as well, and he talked about that.

Why do I get involved in these creative projects? Well, I do it because it's fun, and because it's interesting and I like it . . . and that's the reason I do most of this.

EDWARD COLLINS

Edward is an emerging scientist, a food technologist—still very much engaged in responding to writing and research tasks rather than initiating his own. Much of his research is industry-based, but he has a secondary career as a writer for a board game. Like some of the senior scientists in this volume, he's interested in everything; our broader conversation weaves around living sustainably, photography, and all kinds of writing. He has a gentle, thoughtful air, pausing to think before he answers any question. He has an unusually broad interest in, and experience of, writing—and he sees each of the genres he engages with as influencing the others.

Chapter 6

I Don't Think Scientists Should Be Afraid of Creativity

As well as the science journal type articles, and those kind of things, I do some writing totally outside of the academic area for a game company. In some ways it's not all that dissimilar to some of my academic writing in the process it goes through. We write the material, then it goes through a couple of editions of peer review, and then it gets eventually published. And it takes about two years to go through that whole process.

The game that I write for is set in the year 1220, and it's very concerned about being historically accurate. So it's researching all the things that were around in that period. For example, we wrote a book about cities of that period. So we had to research all the diseases and things that were in the cities, and also the medieval perspective of those things, because, from the point of view of the game, all the things that they thought were true, are true, in that they thought that diseases were caused by demons and bad air. So it's a lot of library type research.

I've always played those kinds of games, and then the company that does this periodically has an open call, where they just get people to submit their own 300 or 400 words about a particular topic. The better ones are then invited to write actual books, rather than snippets of small books. That's how I got into that.

And I do a little bit of creative writing. I write a wee bit of poetry. Nothing much published in the last year or two. Also short stories, which, again, I'm trying to get into a format where I can publish them somewhere. I took a writing course with a creative writer some years ago, and after that I took a freshman creative writing course, and finished my B.A. in English. I'd like to actually do more of the creative writing type material—short story, novel writing, that is what I really want to do. It's a matter of trying to fit it all in.

I think my creative writing has been affected by my technical writing in that I bring a kind of precision to it, that I wouldn't necessarily have otherwise—it's hard to know, of course. And then going the other way, for my technical writing, it's about thinking more clearly about what I'm saying. In creative writing, you're not trying to write exactly what someone else has written, you're not trying to compile a whole heap of what are almost slogans. You are deciding for yourself what you will write. Whereas, looking at some of my colleagues' writing I can see that sometimes they're repeating a few stock phrases—you know, if you're writing a scientific paper, there's a standard way of going about it. So instead, with my technical writing, I'm thinking about it, reading it, asking myself, well, is that really what I want to say? Occasionally, you think "no, I meant to say exactly the opposite to what that actually says!"

Technical and scientific writing are in some ways like creative writing—you start with a blank page, obviously, and sometimes you have to find a new way of explaining something, and that's a creative process. But there are constraints—you have to explain what it actually was that you found in your experiment, so there are limits to how creative you can be. I don't think scientists should be afraid of creativity—some of them might be of course—because science is a creative discipline. Unless you're a technician, it's not about following someone else's recipe; if you're a science researcher, then it's a creative endeavor, and you're thinking of a new question, or a new way to solve an old question, and thinking about what it all means. That is all a creative exercise. Creativity is one of the central aspects of science: it is creativity within constraints. But there are constraints in creative writing too. All creativity is confined in some way.

But the difference is that in science writing, usually you're trying to describe a physical phenomenon, and you're trying to use language to describe it, but the language is not the thing itself, it's a representation of something else. And in science you have to have more fidelity to what it is that you're describing. Whereas, if you were writing a poem, you still have an idea, you still want to describe it in a way that represents that idea, and communicates it to the audience, but you can manipulate what it is you're describing to fit your words. So you have more freedom.

Mostly, in my professional life, I write industry reports. For a lot of the projects I work on, a report tends to be more often the output, rather than a scientific paper, because they often don't want to publish what they've been up to. And other times it's more consultancy than research, so it's often not exciting or novel from a research point of view.

One project I did, which just finished a few months ago, was looking at the energy use in cool stores. The company had had a lot of trouble with the fruit not coming out of storage in the right condition, and so the work that we were doing was just looking at how they've been stored over a season, looking at the energy use and how they could go about saving energy, but not compromising the product at the same time. So the writing for that is like a summary of the findings, for both the company's technical managers and for distribution to the cool store managers, who often don't have a great deal of technical training—they might be the forklift driver who's been promoted up over the years to managing the cool store. So that's part of the problem: the cool stores have often not been run properly because they don't entirely know what's going on from a technical perspective. They've been told how to run it, but they don't understand why. So a lot of the reports are quite graphical—lots of graphs in them, but writing as well.

I don't really get to conceptualise and initiate a project myself—pretty much all the projects I've been involved in have all been something where industry has come saying "we've got this problem, how do we go about solving it?," rather than me coming up with something to do.

Chapter 6

The way they're initiated is that a company will send a description of what they want to a more senior scientist. I think also probably they've talked to him already on the phone, and agreed on what it's going to say before they've sent it. That's something I get to read—then I usually meet with the senior scientist who'd be involved in the process, and then we write some notes about what we're going to do, and then, depending on the extent of it, we might also involve the people from the sponsor (the company commissioning the research) as well. Then we write, just by hand, brainstorming type notes, on what we're going to do there. Then we'd usually write an email document explaining about what we're going to do at various points. Then as I'm doing whatever I'm doing—maybe I'll go out to a site to look at something—then I'll write some notes up afterwards, just for my own benefit, which will then later be used for something else. Then maybe I'm writing a computer programme to simulate something, as part of the project, in which case it would be writing the computer code and trying to document that as I go, so I can explain later to them what was done and why. Perhaps there will also be instructions written for a student or technician who will perform measurements or similar. Then there would usually be one or two PowerPoint presentations that would be done. So I'd be writing those—depending on the size of the project, maybe there'd be one at the beginning about what we're going to do, and then maybe one in the middle and one at the end about what we've actually done. In a small project it would just be one at the end about what we've done. And the written report would go through several hands. Usually I would write the bulk of the report and then it would be edited by the senior scientist in the project, and then often the people at the sponsor would stick their ruler in as well and have a go at editing that.

When I'm doing those industry reports, I'm conscious of the audience. And when I'm writing for the game company, I'm also very conscious of the audience there as well, which is the players of the game. The game has a big emphasis on historical realism—it's not just played by 10- and 12-year-olds, it's played by people in their 30s, or people who have degrees in medieval history, and who, of course, know if you get something wrong. And actually I find that the peer review that goes through is a lot more rigorous than what we do for our academic writing. It is quite interesting how thorough that process often is. Each chapter of the book will get five or six pages of text of peer-review comments about it. Sometimes more if they've found a lot of holes. And then it goes through two peer reviews. There's the first draft which then the line editor reviews internally, and then there's the first proper peer review. And then you revise that, and then there's the second peer review. And then out with it. As well as real content, you're also putting in rules for the game as well, so they have to test that they make sense.

In terms of comparing the two types of writing, with the game company, I'm trying to stick in creative elements as well. You've got to make everything kind of provocative, whereas, with the scientific paper, I'm more concerned with being technically precise. For the game company it's OK to spend a few sentences getting to the point, if it makes for a more interesting read, but the technical scientific writing is usually being written to a stricter word limit, and you've got to get your main scientific points across, with no room for some ambiguous metaphor or allusions. In terms of the game, ambiguity's a good thing. Certainly. Because as well as describing the setting, you're also describing the characters and things that people might encounter, so you've got to talk about what character's motivations etc. might be, for example, but also not dictate everything, because people who get the product want to make up their own stuff. So as the author you can't fully breathe life into the character.

But if we compare technical with the scientific writing? When I'm writing the technical reports, I am thinking usually that I have to explain things in a more simple way and that's because the majority audience for them are often less technical people, so I can't assume so much knowledge (although some of the audience are technical managers who do generally have scientific training and know what we're writing about). So some simple things I would assume the reader of a scientific paper would know or at least if they wouldn't know I can at least put a reference in and say "go read that, it's there." Whereas in the technical one, if I want to make even quite a simple point, I need to explain, walk my way through it a bit more. Simply. In smaller chunks.

I think in the science writing, the blending of having graphs and equations and diagrams in with the writing—I'm not sure whether I can explain it very well—makes a bit difference. Equations and figures can be kind of a shorthand for writing. I put equations in, rather than explaining in words what's going on. Because, in that field, people would understand the equations and wouldn't need the words as well. The figures tell most of the story, even in scientific papers. I think usually the equations are a short hand, rather than a different way of thinking, 'cause you could explain in a paragraph what the equations was saying. It's just that if you understand the equations, they're clearer, shorter, and it's easier to understand some implications.

One of the things I'm never quite sure on is how big the audience for my scientific writing is. If you send an article to a journal, you're never quite sure how many people read those. There would be dozens or hundreds of people in the field, but the matter of how many of them would read every article in that field I don't know. Although even the scientific articles that I write are in an industrial kind of area (refrigeration, energy use and things like that), often quite a wide potential audience will be reading those, I think.

Chapter 6

I think I tend to be a more intuitive writer. I don't usually consciously think what it is I'm changing. I'm just changing it to the right way! Although, having said that, when I'm editing, I do try and scan through and try and get rid of being too verbose in the science writing, which is something that I don't like about a lot of other science papers. I try to cut down on meaningless long Latin words that sneak into a lot of writing. And although you don't use too many metaphors in science writing, if I do have one, then I try to make sure that I've actually thought about what it means, that I have actually used it in the correct sense, so it won't be confusing.

There isn't a big place for metaphor in science. But in the technical reports we might have little colloquialisms (like "capturing the low hanging fruit" when we're talking about what opportunities we're looking at). So there can be little snippets like that, almost clichés, because everyone can understand them. It's important not to try to be too inventive with your metaphors in science writing. There are big commonly agreed metaphors used in science writing, of course, but they are only rarely made up by an individual scientist—they're more like clichés again, such as the "solar system model of an atom." It's no longer the latest thinking, but is still used to communicate some useful ideas.

CATALIZADOR

I drove for five hours in a snowstorm to meet with Catalizador. He is a celebrated chemist whose long career is marked by the highest awards and accolades in his field. He's also a published playwright and poet, is fluent in several languages, and has thought deeply about language and writing. He put aside an afternoon to talk to me about his writing, the connections between scientific and creative writing—and his endeavours to change both the way science is written and perceptions of scientific writing. In this narrative, unlike others, I have retained some of the questions I asked, since they deviated so far from the original question sheet.

POETRY FOR A SCIENTIST IS A WELL-WRITTEN ARTICLE

I read a quote of yours that said "there is metaphor aplenty in science" and yet most of the people I've interviewed have said "there is no place for metaphor in science."

Yeah, they're crazy. They're crazy, but we know where it comes from. Let's blame the current systemisation of science and the privileging of the mathematical, the natural language of science. So that anything that can be made mathematical, converted into an equation of something quantitative, and in further development processed algebraically or geometrically, it is privileged. And those aspects of the human experience which are not mathematical in nature that involve reasoning by analogy, metaphor in all its guises, narrative, telling stories—they're dealt out of science I believe. And that is crazy.

And yet metaphors are consistently used by scientists. Let me give you an example. In my field in chemistry and in physics, we deal with molecules that may have some limited stability; they may be around for a while and so the image, the metaphor that one has constructed for dealing with these involves a landscape with hills and valleys. Molecules rest, for a moment, in valleys far up, are pushed over hills to other molecules. You cannot do chemistry without drawing things, drawing structures.

But the metaphor is of a landscape. It goes something like this; that there is one valley, there's a hill and there is a lower valley on the other side. Molecules move to lower energy, unless something pushes them up. The metaphor is that this is like water facing a dam. The water wants to go downhill, molecule A wants to decompose to molecule B, so what's plotted in the other dimension is the extent of the reaction—from the beginning to an end. This is reactant, this is the product—other very chemical words. And in between there's a hill, and unless there is enough energy to get over that hill, it's going to stay there, like water behind a dam. And most molecules of this world are like that. This paper on which I'm drawing, you know what happens when you put a match to it: it burns. So the molecules in our bodies—the sugars or nucleic which make up the nucleic acids and the genetic material—they're just like sugar, in a way—they'll burn. They are metastable, that's the word to describe a molecule or a substance that is around, but high up in energy. The molecules in our bodies are metastable with respect of burning if there's oxygen around us. But, fortunately for us, there's a hill to be climbed before we start burning.

I use the word hill, but it has to be fancied up in science so what this hill is called in chemical parlance is an activation energy. The expression, fortunately, is pretty plain English: it's the energy necessary to activate those molecules to get over the hill. So how are they activated? By heat of course, or by other sources of energy. And if you make them bounce around and collide, eventually some will climb over the hill. Anyway, here is a landscape metaphor clearly before you—there is a hill to be climbed and there is an energy necessary to overcome the hill. That metaphor has served chemistry now since 1930—very, very well.

Scientists do know they use metaphor. Oh, they do. It's just part of their thinking, because what else do we have? The imagination is very geometrical; we use the metaphors of the world around us—trees are another metaphor. Cyclical things, like seasons, also. So I think scientists know that they're using a metaphor but they don't 'fess up to it.

Chapter 6

Why? Because the metaphor is not mathematical. And they think it won't impress their colleagues. I think if they would admit to using metaphors, the world would be richer, and scientists too, for if they actually talked about using the metaphor they wouldn't be its prisoners. Because in the end the world of molecules differs in some way from that of climbing hills.

So what is a metaphor? A metaphor is a mapping of one part of the human experience onto another one by a mind—that's all it is. All the ways that one has of analogies, simile, metaphor—they're mappings of one type or another. Mappings can never be one to one completely. If you own up to metaphorical thinking, then you can realize its limitations. And look for another metaphor.

As for narrative, storytelling, let me say how I think we come to it. In seeking explanations for things, scientists are no different from other people—they fall for simplicity all the time. It seems to be how our mind is functioning—it wants the world simple. Scientists may think they are smarter than other people—of course, they aren't—and so their falling for simplicity takes various disguises.

So here are three examples of it. There are physicists who will tell you that if the equation that describes a phenomenon is a simple or beautiful equation (they equate beauty with simplicity which is interesting by itself in the context of art. How often would they like to hear Jingle Bells at this time of year—a simple melody—without any variations? They'd get sick in a minute!); then that mathematical equation must be right. And chemists like symmetrical platonic solids like cubes and tetrahedra (because those have a beeline into our soul, or to our minds at least) and have trouble with molecules like haemoglobin which look like a clump of pasta that congealed from primordial soup—there are these chains going every which way in a protein. And my third example of falling for simplicity is that we in chemistry are often looking for mechanisms of reactions. What we mean by a mechanism is enumeration of the elementary steps in a complicated process. So there is an electron knocked off the molecule A, that electron then jumps to molecule B, and B then becomes negatively charged and it reacts with C and it looks like what would be in the UK be called a Heath Robinson-type device; in the United States it's Rube Goldberg (these infernal machines which do a simple task by some complicated manner). So if you give a person a choice between a mechanism for a reaction happening that has one step or a sequence of 37 steps, even if you don't show them any proof, and you ask them which mechanism is right, 90% of chemists will vote for the one step reaction. They're just giving in to simplicity.

So when, through their own hard work, simplicity fails them and the equation is complicated, the molecule looks like haemoglobin instead of a cube, and the mechanism does have, if not 37, then three or four steps, what provides for a scientist pleasure in contemplating the complexity of the real world? I think narrative, telling a story. Even in just saying "electron is taken off molecule A, goes to molecule B, molecule B then reacts with molecule C"—that's a story, it's a narrative. It's got a beginning, it's got an end; it's got process, it's got surprises like climbing hills. You couldn't climb the hill, and then all of a sudden something (a catalyst) comes along and helps you climb that hill.

Scientists often don't realise they're telling a story. This is why actually it's much more interesting to listen to scientists give a talk about their work than to read their papers, providing you're privy to the cognitive structure of the field. And I'm talking about technical talks, chemist to chemist. When they give a talk, scientists relax and they don't have the gatekeepers—editors and reviewers, and their own perception of what a scientific article should look like (that is the worst gate-keeper overall, the one in their head which gets mixed up also with their parents and their teachers and prevents them from writing in good, simple English and narratives). But when scientists give a talk, all of a sudden telling their audience about all those byways and the obstacles, the hill that prevented them, the hero of the monologue, from seeing the solution, becomes a natural process. With one proviso, that the hero always gets over the hill.

So I listen closely to the narrative structure of the seminars; with a little work, I could tell them which Aarne-Thompson type of tale they were recounting. And of course this is so—because the speakers are human, and they're talking to other humans.

I try to fight for narrative and expression and style. In small ways—I can't fight in big ways because I couldn't get my papers published. But I do it in small little ways; I mean it's easier to do in popular journals. But here is a paper that we just submitted to a major chemical journal and I'll just read the first sentence of it.

"Near the bottom (or should it be the top?) of the periodic table, for high atomic numbers, the distinction between valence and core orbitals becomes less well defined."

Well it's that parenthetical phrase—"Or should it be the top?"

I'm talking about the radioactive elements and you could reason—it's sort of weird, I once thought about it, which is why I wrote that sentence—that you're building the periodic table by adding electrons and filling up certain quantised shells. But you're going up in energy: hydrogen is the first and then helium, and the atomic numbers increase. So why are we writing our elements so that the small atomic numbers are on top? Why don't we write it in the way that the energy goes? So that's what that little phrase alludes to. Will I get away with it? We'll see. The reviewers are in the context of finding fault.

I didn't get to chemistry and writing chemistry till graduate school. My personal history is a little unusual in that, while I did play with chemistry sets, I

Chapter 6

did not commit to being a chemist till halfway through my Ph.D. in chemistry. Before that was a sequence of other things: we came as immigrants to the United States when I was 11 and a half; English was my sixth language. It's the only one which I can write, though I speak a few others. So I'm a European from a linguistic point of view—knowing many languages. In high school the only advanced course I didn't take was in chemistry—I took others.

I started the University as a pre-medical student. My father was killed in the war, my mother remarried, and I have a sister who is 17 years younger than I am. We were in difficult straits when I was at the beginning of university. There was a lot of family pressure to go into medicine; part of it comes from being immigrants, part from the family background. In about the first year I decided I didn't want to do that—there was no particular reason, I just wasn't interested. Meanwhile the world was opening up to me in the humanities and the arts; I was exposed to a wonderful art history course. I went on to take some other art history courses and then I took a course in poetry—reading poetry, a great books course—it was wonderful! Literature and art history are what appealed to me.

I didn't think I was good enough for physics. And I think I was scared away from biology by not wanting to be a doctor. I was wrong to think I wasn't good enough for physics because now I work closely with physicists; I just didn't have enough courage. Chemistry was just the natural thing to follow; there were summer research experiences in chemistry that were good; it would satisfy at least my parents' desire that I have a profession—being a chemist counted as having a profession. So I went to graduate school and did chemistry almost because it was the next thing to do and it was easy.

Writing in chemistry, aside from writing lab reports, did not really begin for me until I wrote my first papers. And those happened after about three and a half years of graduate school, at age 25 or 26. I should say that, before that, I was a good writer. I did not try to write any poetry or anything else—I only took a course in reading poetry—but I learned quickly. I remember one instance; in my first year of Columbia I took a required writing course, given by a young English professor. He was sharp, for him I learned some Latin. I wrote a first paper in this class on some subject and it came back—I wrote it in my best secondary school English—full of red marks. It was just terrible. I'd never gotten a C in my life. But I learned.

So we write a lot and my students no doubt model themselves on papers of mine. I worked for my Ph.D. with a professor who had written several hundred papers at that point. It was a large group, there were papers coming out once a month, so there were ten papers a year from the group. That, incidentally, has been the pace at which I have published; not by design, it's just happened. So I have 580 scientific papers published over the years—over 50 years of activity.

The professor drafted the first paper, asked me to add in the computational part and the interpretation. By the third and fourth paper we wrote—and we wrote about six papers based on my Ph.D. work all in one year—I had written the first draft myself. I caught the style.

In my next three years as a junior fellow, I wrote five or six papers all by myself. But then I began to work with a great organic chemist, and we wrote five very important short papers (I was aged 26, 27). He was a master stylist in English; he wrote high English prose and he changed my prose—it wasn't high enough.

I have since gone away from that important interlude. I could not write like he did, I don't want to—I do other things, I use stylistic devices to catch the reader's attention. Of course, I will never write a paper saying "the subject of cesium floride has been investigated in the literature a number of times, references 1 through 7"—how boring to start a paper in this way—I will find other ways to start a paper. One of my stylistic devices is the density of graphic material—that has to do with writing and drawing chemistry. So maybe it's the art that's coming back in some way. Though I think the drawings of molecules are close to the heart of chemistry, because chemical structures are so much part of our articles.

A stylistic device I use is to employ as colloquial a tone as I can get away with, as close as possible to spoken language. I don't use third person passive if I can avoid it; but I sometimes have to fall into that. I will use the present tense rather than the past if I can. Part of the reason for using colloquial language is that I'm dealing with concepts that are rather complicated. So I try to use as simple language as I can. But, once in a while, I slip in a complicated word. It's more likely to be an interesting English word rather than a chemically complicated one. I love it when I can do that. I pull the reader in, then I push him a little. So, aside from telling a story, which is a natural thing to do, I try to get a variation in tempo if I can. It comes naturally, I don't plan these actions, it just comes out that way. I think my little stylistic devices make my papers more enjoyable to read for other scientists. Whether I encounter resistance depends how good I am at it. If I'm good at it they won't know it's being done.

Look at another phrase I've written "we report here, at least in theory;" I could have said "we report here a theoretical study of"—but I don't. I think what I write is more interesting because I hint at "this hasn't been yet done experimentally." And there is another message under the surface of "you can't really trust us theoreticians." Amazing what you can *imply* in a few simple words. Yet I have to do it in a way that will not antagonise the reviewers. It's a balancing act.

The other thing is I don't like hype and extravagant claims; perhaps that's something that comes with age. When you're young you go into the hype mode naturally. You always think that you're the first one to do something and it's the most important thing in the world. And when you're older you smile at that, right?

When I'm writing, I'm in a pedagogic mode. I do want to convince people that this is a good way to look at the world. I know I'm dealing with complicated things; I've got to simplify it, but I'm still talking to my colleagues. But I feel my primary audience is an intelligent graduate student, and not my fellow colleagues, and I've always viewed it that way. I do want to impress my colleagues but I really want to seduce the mind of that graduate student. Why do I want to do that? Well, because I think I have something interesting to say to them; I want them to think, and I know it's hard, it's hard to get there. So I'm trying to teach the graduate student. Sometimes graduate students have told me "your papers are easy to read" or "they're fun to read," and I've said "they're for you." If I write it for the graduate students, some of my colleagues think what I say is simplistic. Indeed, but it's a strategy—I'm trying to make theory so simple that it seems simplistic.

Another thing I've learned (which I do get in trouble with reviewers about)—I've learned a lot from teaching first-year chemistry—is to use what I would call optimal redundancy. So it's the argument of the "Hunting of the Snark"—what I tell you three times is true. You're not supposed to do that in scientific papers, but yet if you really want to convince people, you should say it at least twice—three times is probably too much. So I've got into a lot of trouble because I repeat drawings in papers; I've integrated the drawings into the context of the argument and I feel that if a drawing has appeared six figures before, then I have to bring it to the reader's attention again; he or she will lose the thread of the argument if they have to flick back, refer again to it, you know? I repeat it, but I use stylistic variation to hide repetition. I say it in a slightly different way.

I spoke to someone recently who argued that the words in a physics paper are not important; this person said that he thought in numbers and equations not in words.

I don't think he's analysed really how he's thought. I think he's probably converted the numbers and equations into some furniture of his mind, there are little blocks that he moves around in the mind. I think the intuition is inherently geometrical, drawing on a billiard ball experience. Though some mathematicians have made claims, like this physicist, that they think algebraically. His argument is the standard one and it comes from the idea of universality of science. It's a universality that we aspire to—we really do hope that the antacid I'm going to take in this little pill is going to be the same as that from the next antacid box that I buy. And we really do believe that the design of the factory for making this molecule works in Patagonia and here. There's an interesting story, by the way—the real life reproducibility of scientific findings.

But I don't agree with the person who said the words are not important; I think there is a certain universality, yet there are differences field to field. I would actually make a claim that chemistry is closer to poetry. Maybe because I do both. But I think there are 70 million new compounds that chemists have added to what was in nature, for better or for worse. They have helped to extend our life by a factor of 2. Not for everyone but for many people. And they've caused pollution to the planet as a result of doing the good thing, improving our standard of living and our health. There's a poem by Archie Ammons,

> *Reflective*
> I found a
> weed
> that had a
> mirror in it
> and that
> mirror
> looked in at
> a mirror
> in
> me that
> had a
> weed in it

It is that particular weed, that particular drop of dew that made him write that poem. And the universality was gained from the specificity of the act—that's why you can write a million and one poems about love without exhausting the subject. It's very specific. And $E=mc^2$ is very general—it's a universal statement, for any m. That's one of the differences—one of the relatively few differences between science and poetry.

So I think the person you mentioned is talking about universality, I'm talking about specificity, and there's a lot of that in chemistry and biology, I think. But even aside from that, I think that had he been educated in a different culture, with different teachers, and done his science in a different language he might have come up with a different solution, maybe a solution phrased differently. I'd put it to him that the really interesting thing about the possibility of meeting up with another form of life (which I think eventually will happen, but not in my lifetime) is to see how that different intelligence deals symbolically with the same reality that it and we share; the reality of the universe and its laws. I have no

Chapter 6

doubt that the periodic table will look different or that the equations of physics and chemistry will be written in a different way, even though they describe the same thing.

No doubt $E=mc^2$ would have been discovered if Einstein hadn't done it—so it would have been 30 years later, somebody else would have done it; but what does it matter who did it? And yet it does matter, because the whole shape of the 20th century was shaped by that, by the consequences of quantum mechanics. And perhaps the atomic bomb would not have been thought of in time for the ending of World War II. The world would be different; we'd be worrying about other things.

One person I spoke to pointed to a symbol on their chalkboard and said "when I look at that symbol, it has a whole host of things behind it; it's not just a sentence it's a whole host of things." I didn't think numbers and equations had the same depth of connotations that words do.

Of course. And when someone else looks at that symbol or an equation they wouldn't see the same host of things, they would see a different host. Picasso drew 1,000 bulls, or maybe 10,000, in his very productive lifetime. Each one is different—slightly different. Nelson Goodman, who is a philosopher worried about aesthetics, has written of the difference between art and science. He said that the symbols of art, meaning the mark an artist makes on paper, are replete with meaning, their meaning is not exhausted by one interpretation. I'm not going along (entirely) with deconstruction, that there is no meaning. And that writing (and speech) is the message that abandons, as Derrida would say. The author's meaning matters, but it sets loose a host of other associations. The associations of words or artistic images are endless. I think also in some ways the same things, behind that formula, are let loose by the scientist you just spoke of; he is actually giving a very artistic interpretation of that formula. He should be ready for reading poems because he understands that it isn't just what the word says, it's all the words that sound like that word; it's all the words that are vaguely spelled as that word is. It's all its disparate meanings—all of those resonate in the mind. And give the poem depth and life.

And that applies to numbers and symbols as well. That's why some symbolisms of science are so rich and so persistent and so lasting. Like the periodic table—it's not just the arrangement of the elements, there's much more beyond it. So I find this scientist you quote interesting. And I disagree with you when you suggest a number has the same meaning to everyone around the world. It has the same immediate meaning in the same sense that a dog is a quadruped and not three legged. But then everyone conjures their own kind of dog based on their experience. We must attribute a richness to what's behind the concept of 2.

You wrote that the language of science is a language under stress.

I wrote this in part as one of these things that you write down and then you try to think why it is so. First of all, it's like the man you quoted earlier, our straw man physicist, who would say that the words don't matter. So here is a bunch of scientists and they're talking about things that really matter to them, but the things that matter are in equations and symbols. But equations and symbols, just as facts and measurement of data, are mute; they say nothing. And they say everything, they are very rich. But until a human being works with them they are mute. So the straw man physicist is not going to go to a seminar and just write down an equation without saying a sound; no one would give him a job if he did that. He's going to have to talk about his equation, his discovery. Amusing—there's this person talking about things that matter in the language that he or she doesn't think matters.

The reason I think the language of science is under stress is that our straw man scientist who has the idea that the language doesn't matter is still using language. And people have tried to define things very sharply, the concepts they use, like "force, energy, power, entropy." They have definitions and symbols for them. The language is under stress because those words also had common-sense meanings. In this context entropy's an interesting case, where the word was invented for scientific reasons and has drifted back into common parlance. Other interesting words are relativity and degeneracy. The latter has no moral implications but derives from a certain mathematical use, that two energy levels become of the same energy. So difference is degenerated to similarity. English is a degenerate language. Those words—relativity, degenerate—they have been appropriated by scientists. And then some of the scientific meaning has drifted back into common usage—so I think the word "relativity" cannot be used in English educated speech without some sense of the science behind it.

So the language is under stress—I digress—because there are common-sense language concepts and meanings that are being used to express ideas which people have very precisely defined. But the English words are not so precisely defined, and the language is stressed to make it fit, in some way, the equations and things. Sometimes the words are not there to describe the ideas, yet words must be used.

Let me give you an example which I've used in a poem. I went to a seminar and it was boring, I fell asleep and I woke up and it was still boring. So I defocused from what the speaker was saying and I focused on the language he used. He was solving some mathematical equations subject to some constraint, what is called a "boundary condition." And he said "let us assume free boundaries." Note, two common-sense words used scientifically. Boundary conditions are the limits or limitations on certain variables in equations, that they cannot be bigger than some number or smaller than some numbers—they delineate the conditions for the equation.

Chapter 6

He said "let us use free boundaries." I immediately wrote that down. Free boundaries? That's very interesting. It's inherently poetic because it's an expression of opposites—to me it was. Something was free, but it was also bounded. That was a contradiction, but a very interesting contradiction. He said "free boundaries" and he didn't even think about what he was saying. I wrote it down. And I made a poem out of it.

How did you become interested in writing something other than chemistry?

For me, everything goes back to the university, to reading literature and in particular poetry and plays—I remember a girlfriend dragging me to a play, Garcia Lorca's *Blood Wedding*. I didn't want to go, but then I was just captivated by it. I took a poetry course with a well-known poet. He could not teach writing poetry—that was not done at that time—he taught the reading of modern American poetry. I also took a "great books" course—over a year we read 100 books and that included a good bit of Shakespeare. But I didn't try to write a poem in those university years.

For me the path to writing was slightly anomalous one. I started writing poetry at a certain point, at age 40, and poetry and science were separate worlds for me, though I tried to write about science. The usual progress for people, I observe, exemplified by my late colleague, Carl Sagan, was to do science, being very interested in outreach, writing essays or articles about science, and then writing a novel or a play. Stephen Jay Gould, I think, never crossed over to writing a piece of fiction or a play. Sagan did. But I first had chemistry and poetry, and they were separate parts of my existence. Then I filled in the popular science writing through essays. Then jumped back and I did plays. I have written three or four plays; I've written many essays, some of which have been collected into books.

I write poetry I think because I fell in love with this mode of expression. To keep up some languages that I knew, German and Russian in particular, I sat in on advanced literature courses here—a very important one for me was on the poetry of Boris Pasternak. Whom we know through *Doctor Zhivago*, but who is known in Russia primarily as a poet. And I took a course in German literature, on Goethe's poetry. These courses brought back poetry to me. To me a poem was a wonderfully compact expression of emotions. I wanted to do that.

I thought one could also write poems about science, but that proved to be difficult.

So why is it difficult to write poetry about science?

Oh many reasons. Part would have to do with me, part would have to do with the audience, part with the subject. There is something about science that is inherently prosaic. Part of it has to do with all those footnotes, with worrying about all the exceptions and caveats, where this equation will be right or where it's wrong. I suppose one could turn even these prosaic bits into a certain kind of poetry, and I have sometimes done that—"found poems" in scientific texts. But they're not what poetry's about in general.

To really understand the science, do you have to enter its cognitive structure? There is a wall of jargon around it. I tried in a couple of poems to overcome that. There is a cadence in the language of science, even in the jargon-laden language.

In a poem, I may worry too much about getting the scientific facts right, as if my colleagues were looking over my shoulder, but they're not looking over my shoulder. So if lithium is a solid but not a gas, could I write the gas lithium? If I were a poet one might forgive it, but not really in a chemist.

But there is something else. When you read a poem or when you hear a poem read, you don't understand everything in it, every word. So a disjunction comes about. But usually you can float on the sound, if there is some metre or some rhyme. You float along until you catch meaning again, and then you're with the poet. If too many such disjunctions occur—particularly prone to them are American poems of personal experience where you have no way of making connection with them, with the poet—you lose the reader. The poem means nothing to the reader, after a while. The problem with science is if you use a scientific expression, for example, "the wall of the endoplasmic reticulum," and if the reader/listener doesn't know what that is—or worse, if someone in grade school, let's blame your teacher, gave you the idea that this is science and you must understand it, and if you don't understand it you're stupid—if that train of thought sets in, then I've lost it. So I must work so that this doesn't happen, so that such moments where a person halts and says "what is he saying?" don't occur too often, or don't interrupt the poem.

Shouldn't there be a poetry for scientists?

Most of the poetry of that type—people have tried—it's awful. So poetry for a scientist is a well-written article. That may be the equivalent of a poem. Or an equation—it captures much of reality. In what way? Well the equation is poetic, supposing you're privy to knowing what it means, the symbols and the meanings. It captures the essence of something compactly, economically. It's intense. If you can unfold the equation, it has in it a richness of its meanings. There has been some important work I did where people have told me that reading those papers was like reading a poem.

I think while I write. The longer I've germinated the process in my mind, the easier the writing becomes. It looks like an awful mess when I haven't thought it through much before. I do not write on a computer, there's still a bunch of pencils here—I write in pencil on lined paper. Then I type it up. On the poetry I go through about 10–15 drafts before I type it up; at some point I need to see how it looks on paper.

Chapter 6

One of the joys of writing I think is that you're continually amazed that you have something to say. And this is what's made me a little more sanguine about writing things about anything under the sun—I find that if you will reflect on anything, then the chances are you'll come up with something new to say about it. Because while you may not be original (I'm saying "you," but I mean myself), you are a combination of circumstances of little pieces of knowledge that have been absorbed from teachers and experience and your parents. And the totality is perforce different from that of other people.

CHAPTER 7

"WE HAVE TO COMMUNICATE THE BEAUTY AND THE PASSION."

> I tell my students that you may think you're a scientist—you're not—you're a writer who writes about science.
> — Senior Scientist, Genetics

The story of scientists as writers is a subjugated story, one that hides behind the scientists' overt identity, their central story of themselves as a geologist, physicist, or evolutionary biologist, etc. Many of the scientists in this study struggled to articulate their understanding of writing, because writing—though utterly central to their careers—was something they didn't think about:

> I'm surprised I was able to say anything at all [about writing]. I thought "I have nothing to offer, nothing to say"—I thought it was going to be a real short interview. I think about a lot of things but how I write, the writing process, is not one of them—I just do it . . . I don't analyse it much at all. Isn't that strange when that's the final product of what I do? I mean, people aren't going to come in and say "let me see those parasites now." They want to read what I had to say. (Senior Scientist, Ecology)
>
> This is such a new concept to me. Yeah, I don't think I've ever thought about my writing thoroughly before; you know, you sort of . . . learn how to structure things and grammar and all that sort of thing, but actually sitting down and considering writing is something a bit more than that, I've never done it before. It's been useful. (Emerging Scientist, Exercise Science)

One of the participants referred to this as "an interesting lack of awareness":

> I guess I've never tried to think about this process of my own writing so much before. It's an interesting lack of self-awareness. Because you spend all your time writing basically, spend your time locked to the keyboard . . . if you're not writing for a journal, you're writing a letter of recommendation or you're writing a grant proposal—writing all the time. (Senior Scientist, Biology)

"Are you a writer?" I asked in some interviews, and mostly the answer came back, in some form, "no, I'm just a scientist." Over and over, I noted this odd discrepancy: scientists were concerned about their students' writing, the majority saw themselves as (sometimes reluctant) teachers of writing, but many hadn't recognised the extent to which their professional identity revolved around their writing, until they began to talk about it. Only a few, acknowledging their professional identity as inextricably tied to writing, made writing central to their work with the next generation of scientists.

The story of scientists' perceptions of themselves as writers is also a subjugated story in the literature on scientific writing, a consequence of which has been an insufficiently nuanced picture of the scientific writer focused primarily on the novice/expert (graduate/undergraduate-senior researcher) divide.

And yet, "effective writing comes sometime after the Ph.D." says Timothy in Chapter 3. "Everything I write, I still learn something new. There is always another way of putting an idea, or a fresher way" says Lemrol in Chapter 5. And writing teachers know this—that learning to write is never complete, that there are always new challenges to meet if we are to grow as writers. But if scientists too perceive that writing development occurs post-Ph.D., then investigating how it happens, whether it is successful, and the extent of its influence is likely to be of interest to both the writing and the scientific communities. Furthermore, if scientists perceive that writing activities, beliefs and affect change *predictably* over a scientist's professional lifecycle, or that they may be differentiated in some other way (e.g., by demographics) then this is significant information which emerging scientists need to be aware of and perhaps prepared for. Similarly, if this narrow focus on the novice/expert divide has obfuscated possible variations in writing and identity development, beliefs, attitudes and praxis on either side of the divide (Carter, 1990; Dall'Alba & Sandberg, 2006), then understanding those variations is also important for the ways we prepare our scientists as writers. At the other end of the scale—if writing in school or the undergraduate years is a significant influence on scientists' development as writers—then this should be of interest to the scientific and writing communities alike.

The aim of this last chapter, then, is to develop a broader, and more nuanced story of the scientific writer, and the development of the scientific writer, than is currently available, based on the data set as a whole. Can we distinguish between groups, based on beliefs, attitudes and learning experiences correlated with writing activities or demographic features? And if there are indeed distinctive groups, what are the implications for the scientific community or, more broadly, for those tasked with teaching writing to students within the sciences? How might we, on the basis of the findings here, more effectively enculturate our emerging scientists into their community of practice, as writers of science? The answers to these questions must be approached somewhat tentatively, since the sampling method and sample size mean that the range of participants in this study cannot be seen as representative of the scientific population as a whole. However, the aggregated data may shed light on the development of the scientific writer and the diversity of beliefs and attitudes related to writing as perceived by the scientific community.

LOOKING AT THE DATA

A three-step approach was used to investigate these questions in relation to the entire data set. First, a spreadsheet was used to collate demographic data (including stage—senior, emerging, or doctoral scientist) and writing activities. Second, the interviews were processed quantitatively, based on the model outlined in Chapter 1; each transcript was analysed and scored on a scale between 1–10 for each variable and this was added to the spreadsheet. For some variables (e.g., enjoyment, resilience), the scale simply measured a level, as indicated throughout the interview (where 1 was low and 10 was high). For others, particularly the beliefs, the scale represented sophistication of thought (see Appendix D for more detail on scales for each variable).

The individual scores for each variable were then averaged by gender, professional stage, and designation as adaptive, routine and transitional writers. Finally, to provide more depth of analysis, the qualitative data were coded according to the variables used in the quantitative analysis.

The model used for this analysis, as described in Chapter 1, can be seen in Figure 7.1.

Quadrant 1: Early influences	Quadrant 3: Attitudes
Childhood attitudes/experiences of writing	Enjoyment
Undergraduate attitudes/experiences of writing	Motivation
	Resilience
	Self-efficacy/purpose
Quadrant 2: Learning to write science	**Quadrant 4: Beliefs**
Advisor	Function of writing
Community	Audience
Rhetorical reading	Persuasion
Ongoing support post-Ph.D.	Beliefs about identity/role as a scientist.

Figure 7.1: Model of a scientific writer

Chapter 7

DIFFERENTIATION BY GENDER

No significant differences were found in the quantitative data in any of the four quadrants in terms of gender; women overall did not reveal significantly different attitudes or beliefs, or experience stronger levels of support or more positive or negative early learning related to writing, than their male colleagues. This was somewhat unexpected: although the interview included no direct questions related to gender and most of the female participants did not directly address gender issues, several did suggest there was a gender bias that impacted on their written activities.

These female participants focused on two issues which may impact on their writing. First, several women suggested that there were heavier pressures on women to engage in outreach to schools (as a way of making science more attractive to girls) and/or communicate in a public context—which may impact on a female scientists' peer-reviewed outputs. Second, there were suggestions that female scientists saw and developed their career trajectories differently (see, for example, Bentley & Adamson, 2003). One senior biologist, when told that a previous participant in the study had raised the concept of a lifecycle of a scientific writer laughed "he was a man wasn't he?" Later in her interview, in response to a discussion about a scientist's "genealogy," she suggests a sense of being excluded ("othered"), which may have implications for mentorship and sponsorship of women's writing in science (see Donald, 2013):

> It was another man [who suggested that], wasn't it? Boys do. That's part of the club thing that I would rail against. I don't feel part of that. But I know there are people who do. I find it claustrophobic, the genealogy thing. I think it's true of lots of women; they're not part of that club. And the whole knowing everybody's lineage just immediately "others" you if you're not part of it. (Senior Scientist, Biology)

Conclusions about gender in this study must necessarily be tentative, since the sample of senior women was small. We may hypothesise that women receive less support for writing than men, and given that there is a substantial and rapidly growing body of knowledge (in both the academic domain and in the media) revealing gender bias in selection process and peer review (see, for example, *Nature's* 2013 special edition on Women in Science;[10] Bentley & Adamson, 2003; Dweck, 2006; Hill et al., 2010), such a hypothesis is reasonable. While the quantitative data in this study don't support this hypothesis (figure 7.2), it is notable that the three participants who perceived themselves as being in a sus-

"We have to communicate the beauty and the passion."

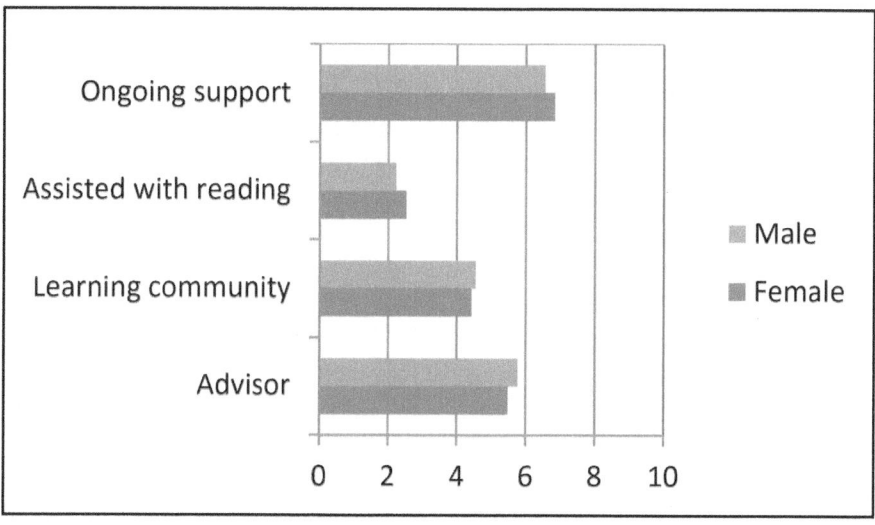

Figure 7.2 Levels of support for scientists as writers by gender

tained, isolated position (both during and following the doctorate) and without community support were all women.

A positive note can, however, be taken away from the gender issues in this study: there were indications that women's support for writing may take a different form from that of their male colleagues. The two writing groups were led by and initiated by women, and were composed primarily of women (Grant, 2006; Grant & Knowles, 2000); both journal clubs (see later in this chapter) were mentioned by female participants. What is encouraging about this finding is that, apart from providing evidence that women are in these ways supporting other women as writers, these initiatives allow writing to be discussed and engaged with explicitly in ways that do not seem to be the norm in the cognitive apprenticeship model.

Given that this study was not designed primarily to investigate gender differences in writing and that the number of women in the sample, particularly at senior scientist level, was small, more research is needed. Fruitful areas of investigation include the impact on women's opportunity to develop as writers of bearing a double burden of being the gender-balanced face of science, gendered differences in career trajectories, and writing support relating to writing (Bentley & Adamson, 2003).

DIFFERENTIATION BY CAREER STAGE

Overall, the data in this study showed support for some aspects of the stages model of differentiation. In particular, two points were agreed by most participants.

The first was that, for most participants in this study, "learning to write science" began in graduate school rather than the undergraduate years. Both the quantitative and qualitative data showed that undergraduate experiences were not significant for developing scientific writing skills for these cohorts. This is perhaps surprising given extensive moves to integrate authentic writing opportunities into undergraduate science education in recent decades through WID and WAC (see, for example, Brieger & Bromley, 2014; Prain, 2006; Reed et al., 2014). While some older participants would have experienced their undergraduate education prior to the impact of WAC, and other participants had been educated in countries where WAC and WID programmes are not prevalent, nevertheless, it might have been expected that many of the younger participants would have experienced some support for scientific writing through their undergraduate years. Those participants who did identify undergraduate education as a significant time in which they learnt to write science again mostly pointed to essay-writing skills, informal writing, or learning to write lab reports—a genre that is dismissed as counter-productive to effective scientific writing by many other participants in this book, a perspective broadly supported by the literature (Driver et al., 1996; Hodson, 1998; Lerner, 2007; Russell, 1991; Russell & Weaver, 2008; Vargas & Hanstedt, 2014; Yore at al., 2002):

> You were taught what a lab report structure was, and aims and methods and stuff [at school], but when I got to doing my Ph.D. I quickly realised that this was just fantasy—there was this myth that lab reports were important, like teaching you for the future! No, it's not! It's not like a scientific paper at all! That's outrageously stupid! I don't even know why we persist with this artifice that lab reports are somehow important. . . . I'd much rather have people fill in boxes with their thoughts that gives them some structure . . . and then later, when it comes to writing papers, they won't have this idea that your paper will be like just a really long lab report. That's just stupid. (Emerging Scientist, Chemistry)

Lerner, (2007, p. 214), commenting on informal writing in the science classroom, suggests:

> Writing in the science[s] often exists in informal modes . . . the kind of writing that is essential for students to do to engage with the material, but not, I would argue, the way for students to learn the relationship between *doing* science and communicating what they are doing . . . And not in a way, in

Russell's (1991) words, "to engage students in the discovery of knowledge, to involve them in the intellectual life of the disciplines" (p.200).

A rare few discussed a specific teacher who required them to write and to think about writing, and only two (Eugene in Chapter 4 and one other senior scientist) experienced anything that resembled a WAC program or writing intensive course. This particular senior scientist (whose undergraduate education took place over 30 years ago in a country where WAC had not been introduced at that time), like Eugene, sees this undergraduate experience as critical to his development as a scientific writer and as facilitating his writing at advanced levels:

> There was a general awareness of teaching writing right across the degree (across years and across courses). . . . But there's one [professor] who was truly inspirational in teaching us how to write. . . . he actually ran a course which was about how you run an experiment and how you report an experiment in written form. It was really a course exploring the philosophy of scientific methodology and an explanation of how you should report science as a scientific paper. He talked about how you write an abstract, what you do put in and what you don't put in, what you can get away with and what you can't. . . . I think the critical step for me was [that] last year of my bachelor's degree. When I went into my Ph.D., I could write science. I was pretty good. I had my skills honed by my Ph.D. advisor, but it wasn't critical. The critical step was earlier. (Senior Scientist, Nutrition and Physiology)

While this study sheds no direct light on the impact of WID or WAC on the development of scientists as writers, there is nevertheless encouragement from these two scientists. Both encountered few difficulties developing their writing skills at advanced levels and both show the attributes of the adaptive writer, suggesting that intervention in the undergraduate years may have a substantial impact on scientists' future development as writers and may counter some of the inherent difficulties in the cognitive apprenticeship model of writing development at graduate level.

The second point that most participants agreed on was that most scientists did experience a significant shift in their writing activities throughout their careers, which appeared to take place 5–8 years post-doctorate (depending on factors such as the length of the post-doc period and employment opportunities)—but further predictable shifts (as proposed by the lifecycle model) were

Chapter 7

not fully confirmed. The predictable move identified by the majority is the shift from writing up their own research to overseeing the research and supporting/editing the writing of their juniors. For some this was a partial shift, but for others it was a substantial change in their writing activities:

> I find now that I write less and less. My job as a writer now is to edit other people's writing. (Senior Scientist, Human Biology)

This shift was often discussed as being, in some way, a surprise the scientist felt ill-prepared for:

> I didn't think I was going to be an English teacher, but that's what I am. (Senior Scientist, Seismology)
>
> I don't think of myself as a teacher of writing, and yet I guess sort of that's what I am doing. I think to myself "I'm a scientist, I'm not an English professor and I really don't know how to teach people how to write." And it's been one of the more frustrating things of my career. (Senior Scientist, Ecology)

The following participant hadn't recognised that she was moving from one mode to another until we discussed the lifecycle model:

> But transitioning through those [stages] is hard . . . because I still expect myself to be—so I'm in that stage of now [where] mostly what I'm doing is working with my graduate students to write up their work. And I think "why don't I value that as much as I value writing from scratch with my own work?" Because I have a lot of my own data that still needs to be written up, but I don't find the time and their work always takes precedence over my own. I should be spending time working on a manuscript of my own, but I don't. So I'm not, I haven't found sort of comf—I feel like what I should be spending my time on is writing up my own work. What I *am* spending my time on is helping others and being at that stage of the life history of the scientist, in which I'm more of a mentor than the person who's putting the work out there on my own. (Senior Scientist, Conservation Ecology)

Some, such as Mason in Chapter 3, reluctant to make this shift, minimised it by allocating as little time possible to supporting the writing of their juniors and retaining a primary focus on conducting and writing up original research. This reluctance was attributable to the value they placed on their own research—that's what got them into science in the first place and it remained a key motivation:

> I hope that I'll always be writing my own research stuff right up to the end. . . . I have a post-doc who spends a lot of her time doing that and it really frustrates me because it's what I like doing too and I'm really jealous. . . . As you become more senior, your colleagues and managers . . . start to ask you to produce a lot more kind of meta-level stuff and it's unavoidable, but I'd hope that I'd always stay involved in the practical stuff as well. Because I enjoy it. Absolutely. That's the thing that gets me into the office; it is still the technical achievement and the curiosity-driven stuff that really gets me going. (Senior Scientist, Interdisciplinary Mathematics)

Others embraced the shift despite the difficulties, seeing it as an opportunity to increase publication rates (Mayrath & Robinson, 2005), develop new skills, and help support the next generation in their discipline. But many clearly struggled for years to feel comfortable with this inevitable transition. Those who could most confidently articulate their role as a supporter of others' writing, such as Lemrol in Chapter 5 and Richard in Chapter 2, were most likely to be close to the end of their careers.

It should be noted, however, that this study focused exclusively on university-based scientists; whether this shift also takes place outside in the academic context (in independent or state-funded research facilities, for example) is not clear.

Beyond these two central points, on which there was almost a consensus, responses to the lifecycle model as a whole were more equivocal. Doctoral and emerging scientists were most likely to agree with the entire model based on how they perceived their own position as writers within the scientific community and their observations of the wider community. But generally the response from the senior scientists was more ambivalent. Some strongly confirmed that this was the model that described their career (see, for example, Marama in Chapter 2) or the way they wanted their career to progress, while others were not so sure or were strongly opposed.

Opposition to the model as a whole by the senior scientists came from a number of positions. Some senior scientists suggested that diversification of writing activities, beyond the shift identified above, was not universal. For example, some senior scientists, as discussed above, did not wish to shift from the position they had established early in their careers as researcher/writers. Other senior scientists had broadened their activities but saw colleagues who didn't:

Chapter 7

> I can also cite very many people in my field who are phenomenally successful who will continue to the bitter end, and beyond, publishing extremely specialised, high-quality papers in their field because they are the leaders of the field and that is what drives them, totally and utterly—[but] I think the truly brilliant chemists are polymaths who can turn their hand to anything. (Senior Scientist, Chemistry)

> I mean there are some people who repeat their Ph.D. research for their whole career. (Senior Scientist, Physics)

Sometimes the move to writing for a broader audience was seen, not as a positive step, but as representing some failure of imagination or intellectual/research capacity:

> I think there's some truth in the idea that you write more broadly as you get older. It might be because you have less time to do research as you get further up in your career. . . . It might be because it's harder to do research—you know there's a tendency to just create a whole lot of stuff in your head so you can't think originally about things. And it can be quite hard to shake off a whole lot of ways of thinking that you become used to. So I think it is generally harder to do research when you've had more experience. But there are positives in that for quite a lot of people, and for me, because you just get opportunities to do things at a broader level. You know, you can influence science policy or government policy. But maybe that's necessity being the mother of invention—I don't know. (Senior Scientist, Physics)

Others, such as Richard and Cameron (Chapter 2), opposed the lifecycle model because they saw it as outmoded and undesirable in terms of the aims of science:

> I would like graduate students, now, to see science communication as an essential part of their youthful role as scientist . . . the face of science is often, I think, perverted by the fact that you see these older people out there who are the famous scientists, and people have this vision of the scientist that way. And in fact science is a youthful game.

Several participants discussed the issue of technology and social media, and how young scientists are more likely to have the technical know-how and famil-

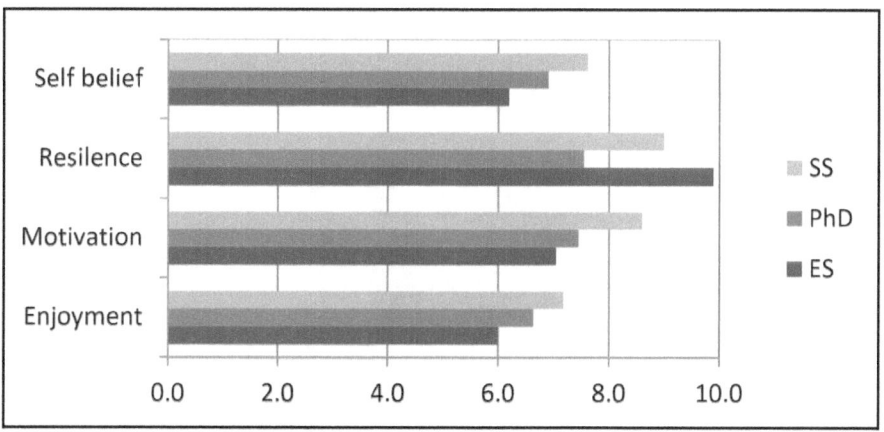

Figure 7.3: Attitudes of senior, emerging, and postgraduate scientists

iarity with social media to reach out to a younger generation—and to be a more attractive face for science.

The results from the direct question in the interviews revealed ambivalence concerning the lifecycle of the scientific writer. But what do the quantitative data tell us—and can we differentiate not just on the basis of praxis but by attitudes and beliefs? To investigate these questions, the data were first analysed by three "stage" categories of participants: senior, emerging and doctoral scientist.

The results suggest that emerging and senior scientists can indeed be differentiated according to attitudes and beliefs. In terms of attitudinal factors—confidence levels, levels of motivation and particularly enjoyment—the senior scientists scored more highly than the emerging scientists (figure 7.3). The only factor on which emerging scientists scored more highly was resilience; as less experienced writers, emerging scientists may be more likely than senior scientists to face criticism and rejection through peer review, and so the need to exhibit resilience may be more important or more uppermost in their minds.

However, the most significant differences between emerging and senior scientists can be seen in beliefs around writing; although none of the rates are high, senior scientists are more likely to perceive writing as knowledge construction and as persuasive and creative, are more cognizant of audience during the writing process, and take a broader perspective on the nature of their role as scientists (figure 7.4).

There was, however, an unexpected finding. According to a stages model (Alexander, 2011a, 2011b; Benner, 2004; Dreyfus, 2005), doctoral students would be expected to exhibit less positive attitudes and less sophisticated beliefs about scientific writing than the emerging scientists. This was not the case: on

Chapter 7

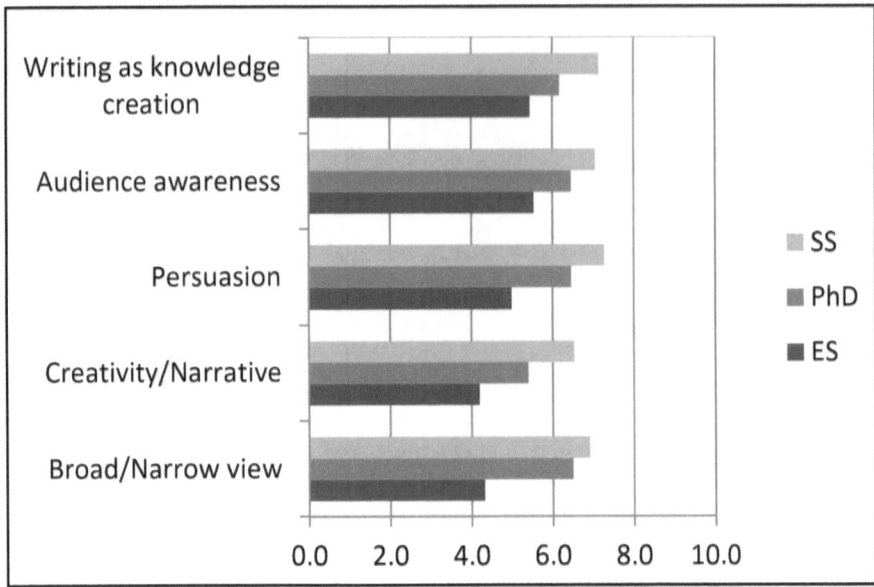

Figure 7.4: Beliefs of senior, emerging and postgraduate scientists

all measures except resilience the doctoral students were closer to the senior scientists than the emerging scientists; they showed, on average, more positive attitudes and exhibited more sophisticated beliefs than the emerging scientists. They were also more cognizant of audience, more likely to see writing as persuasive and to take a broad view of their role as scientists. We will consider this unexpected finding in detail later in this chapter.

DIFFERENTIATING THE SENIOR SCIENTISTS: ROUTINE, TRANSITIONING, AND ADAPTIVE WRITERS

While this study has shown that we can distinguish between the writing behaviours, attitudes and beliefs of emerging and senior scientists, preliminary analysis of the senior scientists revealed this picture to be lacking complexity. Specifically, it didn't account for the fact that the senior scientists exhibited a far wider range of attitudes and beliefs than the emerging or doctoral scientists. Using Holyoak's distinction between routine and adaptive expertise (Chapter 1), the data from the senior scientists was differentiated into three groups: routine (R), adaptive (A) and transitioning scientific writers (A/R).

It should be noted that these three groups did not equate with successful or unsuccessful (or less successful) scientists. Scientists in all three groups could be

"We have to communicate the beauty and the passion."

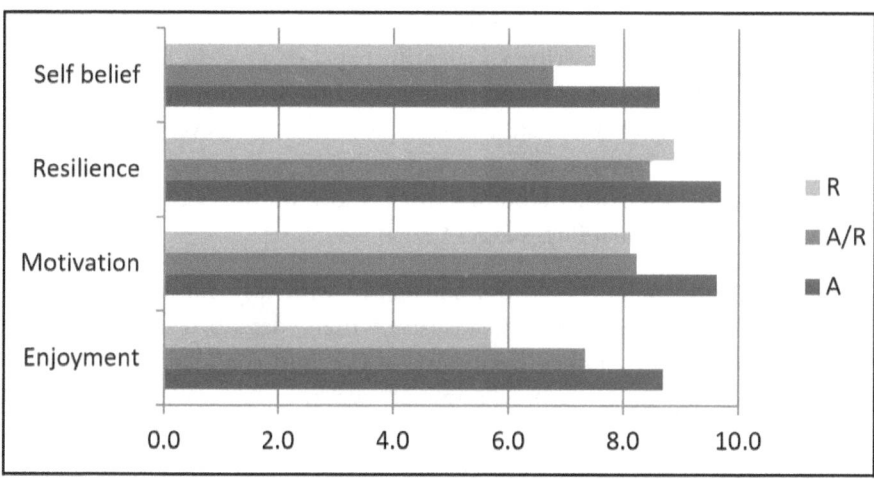

Figure 7.5: Attitudes of senior scientists

equally highly regarded within their professional context. Adaptive scientists did not necessarily, for example, have a more successful career path than the routine scientists, although they were likely to have a higher *public* profile. They did not necessarily publish more, nor were they more likely to have achieved honours in their profession. Adaptive writers published *differently* to the other two groups, i.e., in a wider range of contexts, and they were more likely to be part of the public face of science. In this study, there were equal numbers of routine and adaptive scientific writers (just over 35% in each category), and slightly fewer transitioning writers (29%).

If we examine attitudes and beliefs in relation to these three groups of senior scientists, we find some significant differences. In terms of attitudes (figure 7.5), adaptive writers are far more likely to enjoy writing than routine writers, and are more motivated to write.

Typical comments from adaptive writers included:

> I'd like to write all the time. I'd like to be able to do just writing. I'd like to have the time. And if I have the time, that's what I do. I like trying to write down my opinion about things because that's really hard—because what do I really think? (Senior Scientist, Plant Genetics)

> I think writing, communicating, is very, very important—so I enjoy it. (Senior Scientist, Chemistry)

Chapter 7

> I love writing. . . . And I think there is nothing more satisfying than having written a scientific paper or a review, and getting to the end of it and reading it for the last time and being really satisfied that it's a nice piece of work in every way. . . . I think I'll write till I die! And probably post-science I'll be doing some other kind of writing. I love words! (Senior Scientist, Animal Physiology)

However, enjoyment should not be equated with easy: most of the participants in this study, including the adaptive writers, described struggling with writing in one way or another. Even some of the most confident writers struggled when moving to new cross-disciplinary work, or when writing in a new genre. But, for the adaptive writers, this did not take away from the enjoyment, supporting Bransford's (2004) observation that adaptive experts are willing to engage in the emotional difficulties associated with changing their own thinking and praxis. This quote, from a senior mathematician, seemed to sum up the attitude of the adaptive writers:

> I do enjoy writing. It's extremely difficult for me. (Senior Scientist, Interdisciplinary Mathematics)

While routine writers rated enjoyment low, their scores on self-confidence and resilience were ranked almost as highly as the adaptive writers. The following scientist, who identified five journals he writes for routinely, commented:

> I've done the same thing for 35 years. . . . I'm like a little independent contractor. . . . I've turned down more things than I've probably done, so I do the stuff I want to do. (Senior Scientist, Chemistry)

Throughout his interview, this senior professor described ways of making writing easier and more efficient, in line with Bransford's (2004) description of routine scientists' focus on increased efficiency, discussing the use of boilerplate and keeping to similar structures:

> Writing is easiest for me if it's similar to something I've already done before, because then I can grab it, right. I already have a template for it and I already have some stuff so that will be the easiest.

Both the adaptive and routine writers were more confident about writing than the transitioning writers, which is possibly attributable to the uncertainties of transitioning writers endeavouring to move from one type of writing/research into another. While the difficulties in transitioning from routine to adaptive researcher/writer could be attributed to managing external factors such as other pressing commitments with absolute deadlines (e.g., teaching and administration), internal struggles were often also a factor:

> So my big pause is . . . I think yes, I enjoy it, [but] if I really enjoy it why don't I do more of it—right? [It's like] would I love to get out and hike every day? You bet. Would I like to go for four or five miles instead of my two with my dog every morning? You bet. Why don't I? (Senior Scientist, Conservation Ecology)

This was the dilemma faced by many of the transitioning writers in this study: they enjoyed writing, were strongly motivated to extend the scope of their writing, but the difficulties of knowing *how* to make the transition led to struggle and, sometimes, inertia. The distinguishing factor in many cases between those who had or were making the transition and those who had been poised on the cusp for some time was the availability of a supportive mentor or mentoring context (for example, Richard's encouragement of Cameron in Chapter 2). When ongoing support was coupled with broad beliefs about the purpose of science and the role of the scientist that corresponded with adaptive scientists' beliefs, then the transition appeared to occur more quickly.

Beliefs were critical to the distinction between routine and adaptive scientists (see figure 7.6); adaptive writers' beliefs about writing and their own role as scientists and writers were significantly different to those of routine writers. While adaptive writers show more audience awareness in their writing and are more likely to see writing as persuasive and as knowledge creation, what distinguishes them most are their beliefs about both the purpose of text and their own identity as scientists and writers. Their talk about writing is quite different to that of routine scientists; they talk about crafting a narrative, of shaping the story, of working out where a reader needs what information, of enticing the reader or creating suspense, as this participant describes:

> Good papers are papers that at a certain point show you a glimpse of what is to come, what will be relevant, what is the real explanation of the phenomenon, you know. . . . and to tell this to people means that you have to write a good narrative; if you just rely on the formulas and the words as a glue between formulas and plots it's very likely that your paper will be ignored for a long time. Perhaps forever. [Good scientific writing] takes the reader by the hand; you start a journey

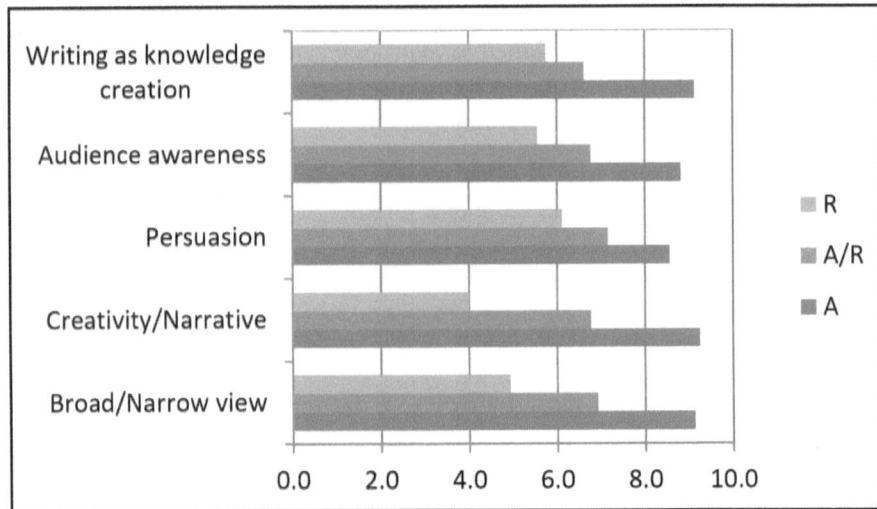

Figure 7.6: Beliefs of senior scientists

together—even if it's a journey through formulas or topics, it is a journey. And you have to organise this journey; and this journey can be pleasant or it can be a pain in the back and that depends on the writing. . . . So I think this is part of the writing process, creating the story. (Senior Scientist, Biological Physics)

Adaptive writers are more able to articulate issues around style, beyond simple descriptors of clarity, concision; they are more likely to talk about voice than simply the use of pronouns, and to be able to discuss how voice changes and why, beyond issues of perceived objectivity. They are likely to show a complex understanding of the interplay between words and figures, and they are more likely to challenge disciplinary conventions:

So there do tend to be constraints, passive voice and so forth, but they're not always followed these days. I wouldn't always follow them. It comes with a kind of maturity, knowing what way you can push the boundaries. (Cameron, Chapter 2)

Routine writers, by contrast, are more likely to see writing as mechanistic, to view writing from a more technical perspective of providing information:

I don't have to be persuasive in my writing. No. I just have to tell facts. But what I do have to be is clear, I have to be a good writer. (Senior Scientist, Chemistry)

"We have to communicate the beauty and the passion."

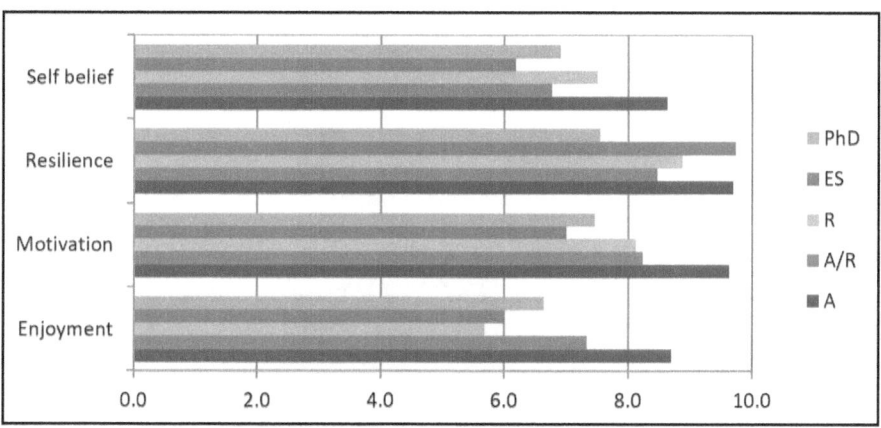

Figure 7.7: Attitudes to writing in all groups

Routine writers stay closer to the model of writing they learned as graduates. For some, the preference would be to dispense with words altogether, and to rely primarily on figures and equations—as they might have done when writing undergraduate lab reports. Routine writers were unlikely to talk about "story"—in fact some saw the notion of story as undesirable:

> And so you have to go back and find the evidence and what does the evidence say? Oh, OK, it actually said this. So when I'm doing strict scientific writing, I keep pulling myself up. I write a sentence and say "oh actually, that's not actually quite right," and reword it according to exactly what the truth . . . what exactly . . . because you get yourself carried away with the story. In fact scientific writing is about presenting the evidence. Not about the story. And I think we've got to be careful not to fall into the trap of telling the story. (Senior Scientist, Plant Genetics)

The transitioning writers were more likely to exhibit similar attitudes and beliefs to the adaptive writers, suggesting either that, in making the move towards a more complex role as a researcher and writer, they must also develop a more complex set of beliefs than they currently hold, or that their attitudes and beliefs are precisely what moves them into a transitioning stage.

If we then put the different groups together—emerging scientists, doctoral scientists and the three groups of senior scientists—we see a more complex picture. The more positive attitudes and complex beliefs of the adaptive senior scientists compared to any other group become much clearer. And while the routine

Chapter 7

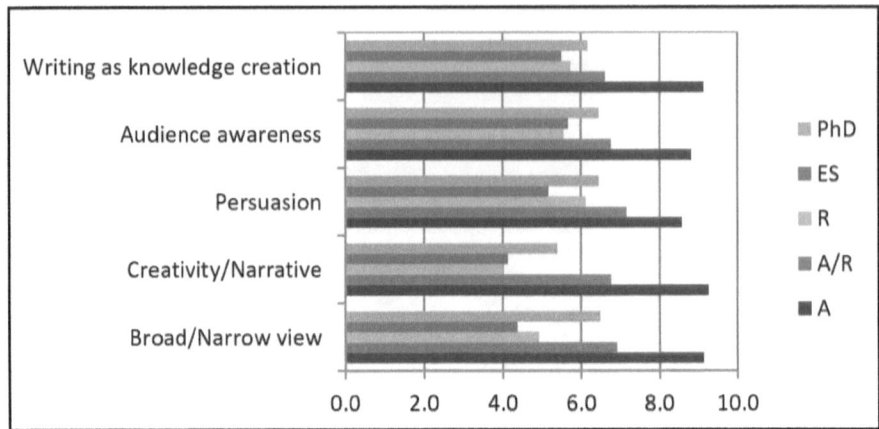

Figure 7.8: Beliefs about writing in all groups

senior scientists remain the most reluctant writers, their scores on self-belief, resilience and motivation remain high in comparison to the other groups.

But in terms of beliefs, while senior scientists clearly have more sophisticated beliefs than any other group, the doctoral scientists' beliefs remain unexpectedly high in relation to every other group.

Setting aside the doctoral students for a moment, what we may be seeing is two distinct yet parallel paths for scientists, predicated on different writing activities corresponding to two sets of beliefs and attitudes: one which follows the lifecycle path hypothesised in Chapter 1 (the adaptive writer), including a sophisticated set of beliefs and positive attitudes, and another path where activities are more consistent (the routine writer), based on more restricted beliefs and less positive attitudes to writing.

An interesting question that therefore arises is whether we can identify those paths in early career, or even early schooling. Do the routine scientists report different attitudes and beliefs in earlier years, and/or did they learn to write differently?

THE GROWTH OF THE SCIENTIFIC WRITER

The results of this study suggest that childhood influences may be associated with the long-term development of scientific writers. The data show that, while none of the groups rated childhood experiences or attitudes very highly, the adaptive senior scientists and the doctoral scientists had the most positive attitudes and experiences of writing in childhood compared with the other groups, while the routine scientists had least positive experiences and attitudes (figure 7.9).

"We have to communicate the beauty and the passion."

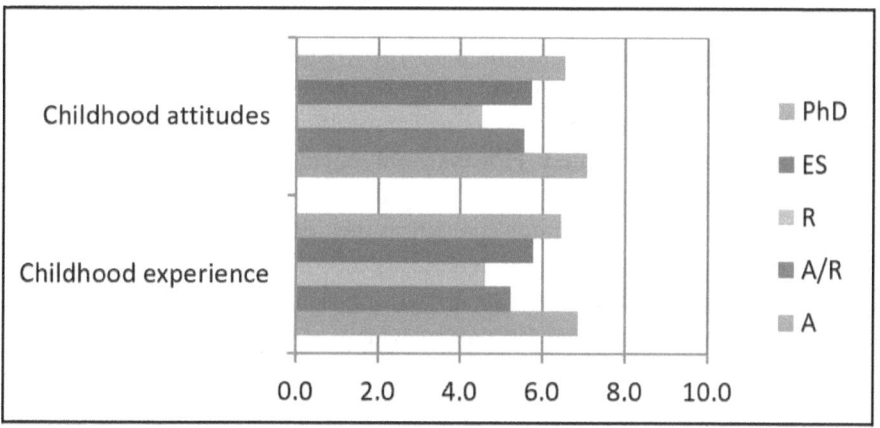

Figure 7.9: Childhood experiences of writing

The routine writers were more likely to have had negative experiences of writing at an early age, most likely to say they struggled with writing in childhood (mostly focusing on the perceived difficulties of creative writing), and most likely to have avoided writing rich subjects in their senior schooling or to see their English teaching as inadequate.

This is not to say that adaptive writers didn't have negative experiences or experience difficulties. Timothy in Chapter 3 is a case in point, but he was not alone—for example, three of the adaptive writers in this study discussed problems with dyslexia at school, and others had had negative experiences with teachers that had undermined their confidence, such as the following:

> I was 8 years old in school in Singapore; I took a cruise with my parents to Hong Kong from Singapore and I wrote a card every day to my class. They were . . . bubbling over with stuff, right, and very messy and, I remember now, the teacher stuck them on the board and said "this is exactly how not to write" because my handwriting was terribly sloppy, so I was obviously creative but I was messy and he made it—he embar—I must have come home and my mother said, "You've stopped writing. What's going on?" (Senior Scientist, Physics)

Some 40 years later, this highly successful adaptive scientist was still apologising to me for the state of his handwriting. But for the adaptive writers these setbacks or negative experiences were overcome—most commonly because they were counter-balanced by some substantial positive support either at school or during the postgraduate years.

Contrary to Martin's (2012) suggestion that scientists are encouraged away from writing-rich subjects at school, most participants *did* take subjects such as English or history right through senior high school, even in countries where taking English (or its equivalent in non-English speaking countries) through senior high school was not required. However, only a small minority said they had learned something about scientific writing in secondary school. Furthermore, those that did suggest that school experiences included the teaching of scientific writing tended to focus on either generic aspects of "good" or "clear" writing, or essay writing (taught in traditionally writing-rich classes, such as history or English), or learning to write lab reports which, as we observed earlier, is not closely related to writing research.

Beyond school, despite the impact of the WAC and WID movements over recent decades, the literature suggests that scientific writing is most commonly learnt indirectly, at post-graduate or doctoral level, through co-authorship, doctoral supervision, and reading and imitation—the cognitive apprenticeship model. This study confirms these factors as important in the development of the scientific writer, but also demonstrates how varied these experiences of learning to write can be for emerging scientists. For some, a doctoral advisor was/is a lifeline in terms of writing and establishing career-long partnerships—for others, not so much. "I think my supervisor might have read my final draft—possibly not" (Senior Scientist, Chemistry) was a sentiment made by more than one participant in this study. For some, immersing themselves in reading in their field made writing a paper feel almost effortless, as if "it just writes itself" (Senior Scientist, Physics); for others it raised questions for which they had no answers and no access to answers. Co-authors' revisions and peer reviewers' comments could be a spur to action or resistance or simply devastating.

More participants had negative or limited experiences of working with a master's or doctoral advisor than those who had positive, constructive experiences. What distinguished most positive experiences was the advisor talking through the changes they were making on a document, but these advisors were a minority. Instead, at doctoral level, if advisors did engage with the writing (and many did not), they were most likely to either rewrite completely, or revise sections without explanation—leaving the student to try to fathom the reasons for the changes for the remainder of the document. This was experienced as highly frustrating:

> I wrote what I thought was an appropriate section in the thesis; it was given to my supervisor and half of it was turned back in red ink as wrong and I felt like I couldn't write anything. So, actually, it was a crisis. . . . I thought I was writing

better and then I suddenly . . . I thought "what on earth is going on?" I was desperate because I knew I had limited time left to finish the thesis and I thought "that's a huge amount of writing I've just done and it's been shot to pieces." (Emerging Scientist, Food Technology)

Even when advisors did talk the student through the process, as this same scientist found, it was often experienced as highly disempowering:

> I know that the thesis was a reflection of my supervisor's way of writing—it wasn't necessarily my style. I queried my supervisors a lot about it, and I fought the content but I let go on the style. They said "this is how you need to explain it" and so I just, I let them—you know, I had to type it in but they were dictating to me, "this is how you write that sentence out." Somehow through that, I actually learnt something. I think it was just because I did not want to have that happen to me a second time, [that] I was determined to understand what they were saying, not just do what they were telling me to. But that was very difficult. They told me that you are supposed to become independent at the writing, and it's like you are a little child again. I had never had to write that kind of way ever in my life; I have never had to. And I do not resent them for it at all, but I wish that wasn't necessary. (Emerging Scientist, Food Technology)

Perhaps the most significant finding in relation to how scientists perceived they had learnt to write in their discipline was that support post-Ph.D. was the most important factor for all groups, including the routine writers (figure 7.10). Writing with co-authors and in ongoing partnerships, often spanning decades (one senior, award-winning physicist noted that he still sends drafts of his papers to his advisor after 30 years—surely the academic equivalent of never leaving home?), was seen by all groups as the most critical for writing development.

For a small minority of participants (all female), writing groups appeared as a support system at this stage. An unexpectedly large number of participants (almost 40%) had attended a writing workshop/seminar during their career, mostly run by an institution they worked/had worked for. However, only two participants saw attending such a workshop as having a significant influence on their scientific writing or writing practices (see Chapter 4), and this was directly related to the development of a writing group. Most participants who had attended writing workshops attributed their lack of value to leadership by

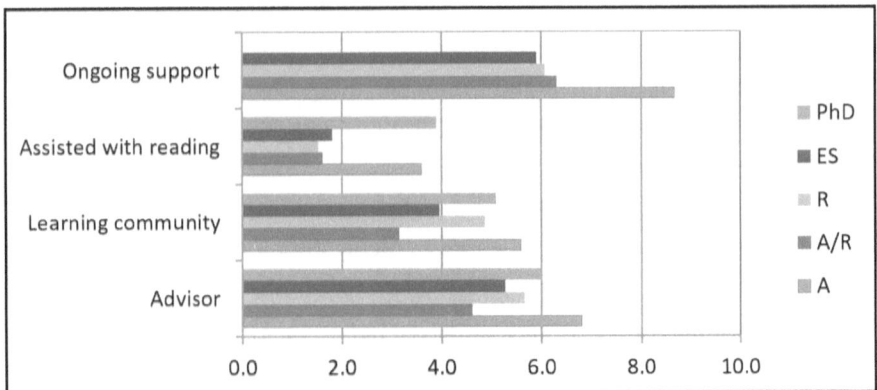

Figure 7.10: Support for writing

people who didn't know about writing in science (see, for example, Wendy's comment in Chapter 3 about the "writing for your doctorate" workshop) or writing in their specific discipline. Positive comments, by contrast, related to convincing empirical evidence that the strategies suggested would work, and opportunities to experience the process of writing science in an authentic setting (O'Gorman et al., 2014).

Mentorship was most commonly discussed as a critical, positive factor post-Ph.D. rather than during the Ph.D. For the emerging scientists, who were at the professional stage of learning to write papers without an advisor, this support was seen as the most significant in comparison to a range of factors during the doctorate.

Adaptive writers and doctoral scientists perceive that they had higher levels of support than the other groups (although the routine scientists do note significant levels of support from their community and a mentor during the doctorate).

Finally, the area where the least amount of support was encountered by all groups was in help with reading. Over a quarter of participants said either no-one had been an influence or a support in their development as writers or that the writings of another author (common influences were "great authors" such as Richard Feynman, Brian Cox, Richard Dawkins and Stephen Gould) were their central influence. A significant proportion of the sample, therefore, was primarily taught to write by reading and imitating the writing of others. I wanted to know what support the scientific writers in this study had experienced in terms of learning to engage with and understand the rhetorical features of writing in their discipline. The answer was: very little.

A few individuals had support in this area. Two emerging scientists, for example, talked about belonging to journal clubs where participants took it in turns to introduce and discuss a journal article. While the primary focus of these clubs was on content, both scientists suggested they had learnt about structure and style while engaged with these discussions. But, generally speaking, scientists in this study experienced a sink-or-swim approach to reading disciplinary texts. For some, this was a successful strategy, but many struggled:

> Because we were never given any instruction or anything or told how to write—it's just a matter of reading journal articles and then trying to [write]; I mean, it was never a conscious thing like I didn't specifically go to an article and think well, right this is how they write so I will go and try and do that; it was just a subconscious way of doing things . . . you get your structure in terms of abstract, introduction, results and so on—and that's basically all you have to go by. (Emerging Scientist, Animal Physiology)

One senior scientist explained this by way of a domestic metaphor:

> My wife is a designer and you know she visually designs things . . . and I look at them and say "I can see that's great because it's simple, it's elegant, it's beautiful, it's succinct visually and yet it conveys the message." But I couldn't create it, I could see it. And the same thing with writing; I can see it in writing but I can't easily create it. (Senior Scientist, Physics)

One of the difficulties with learning to write through unguided reading is that those who manage to this successfully often consequently lack the ability to articulate their understanding to their graduates (Alexander, 2005). The following senior scientist, who writes in three different disciplines, describes how he learnt the different styles:

> You start writing the paper and then you compare it with the other papers [in the new discipline] and there you see [something] that is odd; there is something written in a different way. And then by reading other papers [in the discipline] you absorb the style. The more you read then the more you absorb the style. And actually, for me now it's a kind of a switch in the brain. If I know that it's for a journal I already know, I have a layout of what I have to write in my mind and if I'm in physics the layout is different, you know; the way you order

> your thinking and ideas is different. But then it's something that would be very hard for me to teach to a student. (Senior Scientist, Biological Physics)

A further issue of note is that, while support for reading was low for all groups in this study, adaptive writers were more likely to have experienced such support than routine writers.

IMPLICATIONS OF THIS STUDY

In Chapter 1, I suggested three groups in particular needed to hear the voices of scientists talking about writing: undergraduates, graduate students, and the writing community. But before I turn to these groups, there are a number of implications of interest to the scientific community, and to educational communities in general.

The first is that learning to write science is a lengthy process which occurs from childhood to post-doctoral practice. Furthermore, we limit students' opportunities to develop the capacity, attitudes and beliefs needed to develop as effective writers of science if we focus primarily on the cognitive apprenticeship period. Given the difficulties of shifting beliefs and attitudes (Martinez et al., 2001; White et al., 2005), especially if these are reinforced through poor or damaging educational experiences related to writing in childhood (Brown et al., 2005; Sweeting, 2011; Tapia & Marsh, 2004), the K-12 years may be an important time to embed writing into the curriculum. While some of the participants in this study enjoyed creative writing in the English curriculum, for others it was a damaging experience which either initiated or consolidated beliefs that they were poor writers, that writing was not enjoyable, and that what they learned in English served little purpose and had little connection with their interests and strengths. The only writing any participants remembered within the science curriculum at school—if anything—was the lab report genre which was seen as highly formulaic and requiring no sustained argumentation skills (Lerner, 2007).

While many of the senior scientists in particular experienced their schooling many years ago, and much has changed in our education systems in this time, it is a matter of concern that only one doctoral student and none of the emerging scientists saw school as influential on their disciplinary writing development. Extensive research has been conducted on the impact of integrating writing into the science curriculum at school (see Chapter 1), but the extent to which these initiatives are reaching the students who need them and effecting change is unclear. More efforts are needed to embed writing in the school science curriculum or to broaden the K-12 English curriculum to engage students with scientific interests.

Second, for the majority of participants, the undergraduate years were equally devoid of authentic opportunities to engage as writers of science or to adjust the attitudes and beliefs about writing they had acquired as children. Again, given the impact of WID and WAC, this is somewhat surprising, especially in relation to the doctoral and emerging scientists, and we need to treat the results here with caution, since many of the senior scientists' undergraduate years occurred before the impact of these initiatives. But we might question the extent to which WAC and WID are reaching undergraduate programs and whether they are designed to meet the needs of our science students.

Nevertheless, the doctoral scientists' anomalous scores on almost all variables in our model are grounds for cautious optimism in relation to the impact of secondary and undergraduate education on developing scientists' preparedness for writing. Contrary to a hypothesised stages model of development, they appeared to have more positive attitudes and more sophisticated beliefs about writing and the role of science than any group apart from the adaptive senior scientists. And while their experience of support was low, it was still higher than that of the emerging, transitional or routine scientists on most variables.

As we have already noted, these anomalous findings may be a consequence of the size of the sample or the way the sample was chosen, and so any conclusions must of necessity be tentative. However, we might speculate, on the basis of comparatively high scores related to childhood experiences and attitudes, that changes in the school curriculum related to both science and writing (or English) are having a positive impact. These scientists did not *report* experiences of learning to write science in school or the undergraduate years, but their more positive attitudes and sophisticated beliefs suggest that they have somewhere experienced some effective teaching related to writing and/or scientific writing. Further research with a larger random sample of doctoral scientists is needed to examine whether these findings are generalizable.

What is even more encouraging in relation to undergraduate education is that, for the two participants who had experienced an integrated and authentic writing programme during the undergraduate years, the impact was profound. Both were of the view that, by the time they reached the doctorate, they knew how to read disciplinary texts and how to write. While I have suggested that this study throws little light on the impact of WAC and WID, it does nevertheless confirm the importance of these endeavours, and the difference they can make to those who choose to pursue a career in research science. Perhaps the greatest strengths of WID are its capacity to lay down a foundation of beliefs relating to the relevance of writing to science (the purpose of text and the role of the research scientist, the social context in which science is made, and the role of rhetoric) and to transform attitudes, provide motivation to write, reinforce the

Chapter 7

need for resilience, and restore self-confidence (Driscoll, 2011; Pittam et al., 2009; Rivard, 1994). These attitudes to, and beliefs about, writing are of vital importance in building the scientific writers of the future. Even if academic institutions are not prepared to fully adopt or resource WID programs in the sciences, one of the clear indicators of this research is that some way of engaging undergraduate science students in authentic writing activities would make a great difference to these students' ability to transition into research science.

The third major finding of interest to the scientific community is that the cognitive apprenticeship model, while laudable in theory, in practice is not always all it could be. Sometimes it works. Most of the senior scientists I interviewed saw themselves as teachers of writing, and some showed a sophisticated approach to supporting their students, including an emphasis on enabling the emerging scientist to develop their own style or voice:

> Your job is not to give them the ladder. Your job is to guide them to construct their own ladder. . . . One of the things I say to my students all the time is "my job is not to tell you what to think. It is to help you to understand how to think." In this context, I would say "my job isn't to tell you what to write. My job is to help you develop the mechanisms you will use to write in your own way." And there will be constraints. You may want to write in this way or you may want to write for that audience, and that way may not be appropriate. So you have to learn how to write for that audience. . . . the job of the supervisor and the other people—effectively scientists in this context—working with their colleagues as well as students is to quite consciously draw out the mechanisms that they use, and in this case to build their own ladder, and the more rungs you have on the ladder the more you can push forward. (Lemrol, Chapter 5)

One of the most difficult experiences of the doctoral students (or those remembering their doctoral experience) was uncertainty about asking for explanation from someone who is extremely busy and more senior—and who they experience as highly critical. Yet most of the senior scientists I interviewed felt a strong sense of care for their students, and sometimes the problem was really just part of the muddle of human relationships:

> One of my Ph.D. students told me two or three weeks ago . . . that I didn't care about the project anymore; I wasn't interested. And I said "of course I am—I'm just really busy."

> And then I got an email from that student on Thursday or Friday last week . . . and she said "I wish I'd not said that to you." And I just replied and said "I'm really glad you did because you need to make me aware of that, and you also need to remember—well for me at least—you're like my family. You're the most important part of my working life. You're the ones that keep me going; you're the ones that make the new compound." I mean, they've given me the buzz and the drive and I feed off their enthusiasm. And I also said "you've got to remember that, like my family, you see me when I'm most tired, most grumpy; you get the poorest quality time of everyone else. And that's wrong and I'm sorry, but that's how it is." (Senior Scientist, Chemistry)

But problems remain around the doctorate. Part of the problem lies simply in the uncertainty around the advisors' capability and willingness to engage with their doctoral students' writing. It seemed some students were lucky and others were not:

> I think our system of teaching graduate students to write is pretty bad because there is nothing that is really implemented and it's all left to the individual, so it's a bit of luck. If you come into a group where there is a supervisor that is caring, you get some support. If you go into another group you get absolutely nothing. (Senior Scientist, Chemistry)

For some there were ways around deficiencies in their advisor's capacity or willingness to engage: some called on support from their lab partners (e.g., a more experienced student, like Jane in Chapter 3), others found another mentor outside of the doctoral process, while others called on family members (e.g., parents or partners—see Grace in Chapter 5) or friends, some of whom were scientists and others "just good writers." But the fact that over a quarter of participants could identify no form of support, and that many others who did experience support had no help in interpreting revisions, remains a matter of concern.

And it was a matter of concern to the advisors too (Maher et al., 2013). Many of the participants in this study struggled in their role as a teacher of writing: "I'm not an English teacher!" was a common . . . no, not complaint, more like a heartfelt cry. While some, such as Richard and Lemrol, had a firm grasp on this role, others struggled with this turn in their careers, feeling ill-equipped and ill-prepared. And we could see this as a general problem of doctoral supervision—whatever our discipline, we may all struggle with learning the strategies of graduate advising—but I would argue that the problems are more acute and have greater impact in the sciences (Gardner, 2007, 2010; Golde, 2005, 2010; Rodriguez et al., 2012). It would be a rare humanities scholar who gave up writing their own research, and whose writing then primarily constituted forming co-authorship relationships in which their role is editing the work of others. Yet this is precisely the career turn that many university-based scientists seem to take (Maher et al., 2013), not many years past the submission of their own doctorate. The difficulties of engaging in this role (Lee, 2008; Paré, 2011) therefore have a career impact on both the advisor and the emerging scientists with whom they are working. Emerging scientists' difficulties with writing have implications for the advisor's career—a situation that is not replicated (or at least, not to the same extent) in other parts of the academy.

So, are standalone workshops on scientific writing the way forward? The problem is credibility, of writing being validated by someone who is perceived as credible (see Richard, Chapter 2), and socialisation (Brown et al., 2005; Ding, 2008). WAC and WID instructors are aware of the issues of authenticity and socialisation, but since WAC and WID course are perhaps most often focused at undergraduate level, we may be missing opportunities to grow our emerging scientists as writers in these final years of formal education. Instead, my suggestion is that we need equip scientists—and students—with an ability to articulate their rhetorical decisions (Beaufort, 2004). They need a language with which to *talk* about writing. And perhaps this is where the writing community can be of most use. Perhaps collaboration between science and writing faculty through WID is the best way to ease this pathway, as long as we are prepared to genuinely listen to one another and if we continue that collaboration further into doctoral advising. Whatever method is used, this study suggests that the scientific community would benefit from examining the way scientists are prepared to become teachers of writing. The current system is too unreliable for both the senior scientists, who often exhibit considerable anxiety around mentorship in relation to writing, and for the emerging scientists they are advising and with whom they are engaged in co-authorship.

The fourth significant finding of this study is the possibility that, beyond the predictable turn in most academic scientists' careers from writer to teacher of writing, scientists make choices about remaining on a narrowly focused disciplinary path or broadening their scope as both scientists and writers of science: the routine and adaptive pathways. Furthermore, writing (and differences in attitudes to and beliefs about writing, their role as a scientist, and the purpose of science) is central to the choices they make. It is clear that routine and adaptive scientific writers have equally valid career paths, and both are essential to the progression of science. But if we feel that scientists' engagement with public discourse around science is important, and that more scientists need to engage with non-scientific audiences (Brownell at al., 2013; Greenwood & Riordan, 2001; Leshner, 2003; Olson, 2009), then preparing emerging scientists for an adaptive role may mean working with their beliefs and attitudes, perhaps from an early age, and providing more support for them as writers.

What this study has not been able to achieve is an answer to the question of causation. While we have been able to show that routine writers are more likely to have negative experiences of and attitudes to writing as children, there is no clear causal link—indeed, some scientists with very negative childhood experiences of writing went on, with ongoing and positive support, to become high-profile adaptive writers. Similarly, it is not possible to ascertain whether different attitudes and beliefs caused or were a consequence of the choice to become an adaptive or routine writer (Chamberlin, 2010; Sweeting, 2011). Some participants in this study were able to identify the impetus to broaden their audience—some wished to kick start a new direction in their career, others were ambitious for research funding and saw reaching out to cross disciplinary teams as a way to achieve this; others were encouraged by senior mentors or advisors. Further research is needed if we are to understand these choices and develop pedagogy to address them. In the meantime, the challenge for the scientific community is to begin to model and articulate the attitudes and beliefs towards writing they wish to see in their graduate students.

IMPLICATIONS FOR THE WRITING COMMUNITY

There are many possible implications from this study for the writing community, not least of which are the significance of the undergraduate years, the value of engaging with attitudes and beliefs, and the importance of seeing the scientific community as diverse in ways that go beyond disciplinary differences. However, I want to focus on just two issues here.

The first issue is a question of trust—and listening. A comment in an early WAC text has stayed with me throughout my career of working as a teacher and researcher in the sciences:

> [The engineering professor] fixed me with his eye and went on to say that we across the disciplines don't trust each other. I think he was right. My experience [of WAC] has, however, illustrated for me my own biases and lapses of trust. I am more aware that we college instructors . . . walk a common ground. Unfortunately, this common ground is often

> obscured by disciplinary boundaries and professional loyalty. (Yancey & Huot, 1997, p. 77)

Mya Poe, Neal Lerner, and Jennifer Craig (2010, p.199) end their book with a frank discussion of the relationships built through their WID program, commenting on the learning and frustration of engaging in a collaborative relationship with STEM faculty:

> Such reciprocity is rare in academia, where faculty often stay in their respective disciplinary territories and only occasionally venture to the other side of the campus for day-long teaching workshops. At times our collaborations have been frustrating, as our values and background knowledge seem so disparate. Yet with each iteration of our work we have come to better understand and appreciate the varied perspectives we both bring. . . .

The need for effective dialogue is generally seen as the cornerstone of collaboration—Brammer et al., 2008, for example, outline the history of conflict in cross-disciplinary pedagogical collaborations and pick up Pratt's (1991) concept of the "contact zone" as a way of finding common ground. "We must," they argue, "resist both colonizing and bring colonized," suggesting that "real" dialogue about learning to write will reveal, between writing teachers and faculty from other disciplines, more commonalities than differences.

But I would argue something quite different. While common ground is important, there is danger, in searching for those places where we meet, of failing to realize that the language we share (such as collaborative writing and peer review) may, in our different cultural contexts, means quite different things. I would also argue, since the research to date into scientists' perspectives of writing has primarily differentiated the scientific writer only by discipline or by a novice/expert divide and failed to see the immense complexity of the scientific community, that true dialogue has been hard to find.

Perhaps we need a new word that takes us beyond "dialogue." In Maori, the word *kōrero* (both a verb and a noun) means conversation, but a particular kind of conversation that prioritizes both respectful listening and authentic engagement (Soliday, 2011). I would suggest that it is by engaging in the *kōrero* that we become less fearful of the differences between us; that in the *kōrero*, instead of seeking out commonality, we will see differences more clearly and perceive those differences not as barriers but as energizing and creative opportunities. In this process, we will begin to see an altogether more differentiated story of scientists as writers—a story that has exciting possibilities for collaboration and pedagogy.

"We have to communicate the beauty and the passion."

How do we engage in this way? There are some obvious starting points, such as taking time to ask questions of our STEM colleagues and listening for differences instead of commonalities, listening for the *taonga,* the unexpected story. But perhaps there is also a role for the writing community beyond its current limitations of collaborating with science faculty through WID and WAC during the period of formal education. Perhaps there is a place for writing teachers within WAC or WID programs to work alongside scientific research communities to help articulate, *within research teams,* the role of writing in research, and to provide greater access to language that would make writing more visible, more conscious, more manifest (Artemeva, 1998).

Three significant outcomes would emerge from this: first, emerging scientists may have more direct access to conversations about disciplinary writing, which would ease their transition to becoming writers within their discipline. Also if, as some of the participants in this study suggested, communicating across disciplines and to the broader public, is a moral, social, and economic imperative of science, then integrating writing teachers within research teams could provide a transitional non-scientific audience and a language and incentive to invite the "deep" conversations about the relationship between writing and science, and could thus enable existing learning communities in science to become expert writing communities (Bereiter & Scardamalia, 1993), extending their science beyond their disciplinary boundaries. But most importantly of all, writing teachers would gain from a more authentic engagement with, listening to, and understanding of scientists writing; much as we have to contribute, we have more to gain (Brammer et al., 2008; Mullin, 2008; Segal et al., 1998).

IMPLICATIONS FOR STUDENTS

There are multiple implications from this research for both undergraduate and graduate students in STEM. We might point to the broader findings related to the significance of writing, and attitudes and beliefs about writing, in relation to career direction. Or we might focus on the need to find an advisor or mentor who can talk about writing, the importance of learning to read not just for content but for the structural and stylistic attributes of a text, or the value of perceiving learning to write as a more intentional part of the cognitive apprenticeship process.

But, instead, in these last pages, I'd like to point to something else. My hope is that the students reading this book will focus on the particular experiences of the scientists who tell their story as writers of science. I hope they will be encouraged into resilience by reading of the struggles of the successful senior scientists, some of whom have overcome significant obstacles to become scientific writers:

> I was not born writing the way I write today, I had to struggle and learn exactly what you have to learn. (Lemrol, Chapter 5)

> I wouldn't say I'm a good writer now. I write terrible sentences and I have to correct them. I know how to do that. But, you ask me to write a paragraph and I write it down and then I read it back and I think "oh my God!" It seems OK as it's coming out of my head, but it's not OK when you read it . . . It's never gone away; I never find it easy. . . . But I do have a revision facility. (Timothy, Chapter 3)

> I remember taking an English class at university and I was lost. I was like, "I don't know what any of this stuff is." I just did not come from a really strong place and it's literally just been practice and good advice. And lots and lots of support. (Lizzie, Chapter 5)

I hope that, in reading these stories, they will see behind the confident exterior of their professors to the emerging scientist they once were, struggling with dyslexia, puzzling over the enigmatic scribblings of their advisor, staying home to write in their pajamas. They too found—and often continue to find—writing hard.

I hope the students reading this text will have learnt something about managing the gatekeepers, of knowing when to pull themselves together and learn from critique and when and how to resist (Lemrol, Chapter 5; Catalizador, Chapter 6). I hope they will have been empowered to articulate to their advisors what they need as writers: that they need not just corrections or revisions, but someone to talk through the revisions so that they can make more intentional and conscious decisions about writing; that they need a learning community that can engage with uncertainty about writing.

I hope they have learned something about writing process, particularly about writing from the inside of the paper outwards and the integral structure of a scientific paper:

> What I think young writers do is they write their intro first and it becomes this big blathering of everything they've learned on the topic and that's not what a journal wants to hear and it's not what a reader wants to hear. . . . you should write your results first, and then write the methods that explain your results. You don't write everything you did—you write only the methods that explain what you want to talk about. Then you write your discussion, which doesn't talk

> about anything except the results that you presented, and then you go back and write the intro that would lead you into that. That's very different from writing the intro by talking about the subject, especially when you're given the task of "do a literature review on everything that has to do with your topic." (Eugene, Chapter 4)

> [T]he whole structure of the paper is constructed around the hypothesis and the aims. So for example, let us say you have the hypothesis "if x then y." In writing the introduction you have to introduce x, you have to introduce y and you have to introduce the relationship between the two. Your hypothesis "if x then y," also determines your aims. The aims then automatically determine your materials and methods. And the aims then also automatically determine the laying out of the results. And the relationship between x and y automatically determines the structure of the discussion. (Elizabeth, Chapter 4)

I hope they've learned something about the centrality of the story and how that story can emerge from the "pictures" (as Richard discusses in Chapter 2) or from talking with others (Gao in Chapter 4).

And I hope they've learned something about agency, about how, in the absence of an effective mentor, they may seek out help from lab partners (Jane in Chapter 3), family members (Grace in Chapter 5), and friends; that they'll feel empowered to set up journal clubs and writing groups (Elizabeth and Sally, Chapter 4) and to build a supportive learning community around them. That they will understand that such groups do not need expert leadership, they simply need a group of peers who are willing to work with uncertainty together.

Perhaps most of all, I hope the students reading this book will pick up Richard's claim that "science is a youthful game," that they have a crucial role to play in both moving science forward and in engaging with the public understanding of science. I return to Randy Olson's comment (p.8) about the public's fear and rejection of science:

> It is a genuine threat to society. In the midst of this conflict, communication is not just one element in the struggle to make science relevant. It is *the* central element.

The hope of many of the scientists in this study lies with the young scientists who are emerging out of our classrooms and laboratories today:

Chapter 7

> It is vital that these young people are able to think about these questions, think about context, think not just about the economic value but about the human consequences, the ethical issues that sometimes arise in science. Also to communicate the beauty and the passion around the subject and get people excited. So they see that science is . . . it's a wonderful thing. (Richard, Chapter 2)

LIMITATIONS OF THE STUDY

I was reminded, as I reread the interviews towards the end of this study, of the prevailing notion of "story" in this text, and Catalizador's observation that we are all engaged in writing stories:

> But when scientists give a talk, all of a sudden telling their audience about all those byways and the obstacles, the hill that prevented them, the hero of the monologue, from seeing the solution, becomes a natural process. With one proviso, that the hero always gets over the hill. So I listen closely to the narrative structure of the seminars; with a little work, I could tell them which Aarne-Thompson type of tale they were recounting. And of course this is so—because the speakers are human, and they're talking to other humans.

The primary aim of this research has been to offer a *taonga*, to tell a story, to invite the reader into the *kōrero*, into an engagement, at once empathetic and critical, with a series of voices that tell of scientists' experience of writing science—and to provide more insight into the diversity of how scientists see the relationship between writing and science.

There are, however, limitations to this study in terms of its secondary aim of providing a nuanced story through story analysis. While the sample size was larger than other studies that have engaged with scientists' perceptions of science, it still has limited scope, and the purposeful sampling technique means that the sample cannot be considered representative. It is highly likely, for example, that a more random sample would show a higher overall proportion of routine writers; this is reinforced by Yore and associates' work which clearly defines scientists' perceptions of writing from the point of view of the routine scientific writer. A larger sample of doctoral scientists is needed to give a clearer perspective on whether the changes indicated in this sample are more broadly applicable or how extensively changes in the curriculum and WID programs are impacting on science education. A larger sample of female scientists is needed to show whether women experience different levels of support as writers and whether they are supporting each other differently. A study which intentionally collected data from a limited number of defined disciplines would allow for a comparative analysis across disciplines.

The data collection methods used in this study present a retrospective picture of the scientist as writer. Apart from issues of reliability, which were discussed in Chapter 1, some of the experiences of the senior scientists—their experience of their education, for example—may be no longer applicable in a changing educational, social, and technological context. This study has addressed this issue by using a varied sample, so that the experiences of younger scientists (emerging and doctoral) can be compared with those of the senior scientists. A longitudinal study may provide richer and more reliable results, but this presents obvious challenges and will be someone else's work.

The data collected during this study is more extensive than could be discussed in one text. Numerous other themes could have been pursued, such as more in-depth discussions of scientists' struggle with the word "persuasion," their writing process, how the way they read a scientific paper often contradicted their explanation of why they wrote the way they did (Bazerman, 1985), how they perceived the relationship between words and figures, and perceptions of the scientific paper as primarily pedagogical. In this text, these themes were simply part of the context of a more focused analysis.

THE *TAONGA*: "WE HAVE TO COMMUNICATE THE BEAUTY AND THE PASSION"

Three unexpected words recur through the interviews in relation to writing science. The first two were "beauty," and its close cousin, "creative." While several participants deplored the quality and ugliness of scientific writing, many other participants talked about its capacity for beauty. Catalizador is somewhat disparaging of scientists' views of beauty as simplicity, while at the same time seeing the equation as fundamentally poetic. But simplicity was not what constituted beauty for most of the interviewees who discussed this. Instead they were likely to talk about clarity, accuracy, and effective story-telling, about shaping a narrative creatively so that it would engage rather than simply inform a reader.

The other unexpected word was "fun." "Why did you choose to write across disciplines/take up blogging/write poetry?" I would ask—and again and again, the response came back: because it's fun. I was left with the enduring sense that many scientists—in particular, the adaptive scientists—enjoy risk when it comes to writing. Often I asked the *why* question when someone had just finished outlining in detail the trauma of learning a new way of writing or of engaging with a new audience, or of feeling exposed in new ways. In fact, I can think of only one participant who actually suggested writing was easy; for many it was hard . . . but fun.

In thinking over this whole study, I keep coming back to Gao's comment about his choice of discipline: "I learned in my graduate programme that you're supposed to identify hard problems to work on and you're supposed to take your creativity and solve hard problems." It seemed to me that writing—and continually choosing to develop writing in new ways—was, for many of the scientists in this book, about taking their creativity to work on "a hard problem," but one that most of them actively chose to pursue, at least in part, because it was creative and fun. Many—perhaps most—are poorly prepared for this hard task, and their apprenticeship for writing science is intrinsically and intuitively intertwined with the disciplinary apprenticeship they encounter from graduate school onwards. Yet, as they engage with a changing social context for science, they are engaging as individuals and as communities, in the hard task of writing science.

AFTERWORD

My research has taken me around the world in search of scientists' voices. On the way, I have been assisted by many people. My first thanks go to all the people who volunteered or agreed to be interviewed for this research: it was, indeed, a privilege to *kōrero* with you. A special thanks to those whose extended narratives appear in this text: you are all busy people and I was honoured that you put aside the time not just to be interviewed, but to revise and edit and discuss the narrative.

I'd like to thank the team at the WAC Clearinghouse, Mike Palmquist, Sue McLeod, and Rich Rice, for their enthusiasm for this book, their encouragement, and for the way they have patiently guided me through the process.

I am grateful to my home institution, Massey University, for providing support in many ways across the years—in particular I would like to thank John Muirhead, who found money to help and time to discuss my ideas. My thanks go to Iain Hay, and Mark Israel, both then of Flinders University in South Australia, and Sally Mitchell of the *Thinking Writing* programme at Queen Mary University London, who hosted me during periods of data collection and were generous with their time and their ideas.

Thanks are due to Suzanne Lane, Director of MIT's WAC program, and Neal Lerner from Northeastern University in Boston—and a particular thanks to Neal who made a chance remark that made me see the structure of this book whole. Warmest thanks to Susan Ruff, of MIT, whose remarkable work with mathematicians opened doors for me, who shared ideas and interpreted the mathematics classroom for me—and who didn't give up on New Zealand even when torrential rain trapped her in a hut on a mountain climbing expedition.

To my colleagues at the University of Vermont, and in particular the folk in the WID program, I am most deeply indebted, both personally and professionally. Susanmarie Harrington and Ellen Andersen, Sue Dinitz, Kristen Cameron, Sharon Henry, Kathy Fox, and their families, were the warmest hosts during my family's five-month sojourn to the US. They opened their homes and shared friendship, ideas, and laughter in equal measure. I hope I have been able to offer something back in small part for their warm generosity and hospitality.

To my transcriber, Lynn Hyde, I offer my heartfelt thanks. Lynn took a keen interest in her work, and her comments often helped me to see things in new ways. I hope this book will somehow compensate her for the hundreds of hours she gave up to sit and listen and type while the sun was shining and she wanted to be out on the farm. Thank you to Rose O'Connor, Jenah Shaw—*you amazing girls!*—and Rahna Carusi who all assisted with transcription.

I'm also grateful to the friends who have helped in so many ways, especially Sue Fordyce and Robyn and Glenn Mason, who offered me places to write when I desperately needed to concentrate, and Anne van Gend who picked me up and dusted me off at critical moments. Thank you to Jan Dewar, Elizabeth Schaw, Jane Wilkinson, and Esther Garland, who cheered me on, Jordan Massicks for helping me build muscle—and to Anna Greenhow, Neil Bruere and Hugh Kemp, who took such a lively interest in my work and offered resources and ideas which invigorated my teaching and thinking. I would like to acknowledge a particular debt to Pete McGregor whose insights, offered from the perspective of someone who is a friend, a scientist and a writer, were invaluable.

I am fortunate, too, to work with a remarkable group of people. Thank you to the dream team, Angela Feekery and Ken Kilpin, Anne Meredith (without whom nothing would be possible), and my wonderful tutors: Judith Moore, Megan Stace-Davies, Louise Folster, Jacqui Burns, and Jo Vitkovitch. Thank you for taking on the weight of our shared teaching when my mind was completely occupied by this book.

My thanks are due, most of all, to my family, Eddie, Rose, Emily and Lizzy, and to my parents, Jean and Ellis, and sister, Anne-Marie, who have supported me in a myriad of ways throughout the writing of this book, and who shared the adventure of travelling the world to collect the data that makes up this text. And to my husband, Bruce, who crunched the numbers, drew my graphs, took over the household, and kept me on track even when he was up to his ears in his own work, thank you is not enough—but, *thank you*.

Finally, I would like to acknowledge Fulbright New Zealand. I was fortunate to be awarded a Fulbright Senior Scholar Award in 2012/13 to pursue this research, and without this support, the interviews could not have been completed. One of the aims of the Fulbright Foundation is to promote mutual understanding among cultures—and that is one of the aims of this book, to promote understanding between scholars and teachers of rhetoric and composition and scientists. One of the participants in this book in this book, talking about his work as a chemist and an educator, describes himself as a *translator*. The metaphor I would use for this book is that it aims to be a bridge. I would hope for this book to provide a pathway across C. P. Snow's infamous gulf, so that scientists and writing teachers and scholars can meet as equals in their shared concern—and their differences—as writers and teachers of writing.

NOTES

1. The use of colonial language is intentional. It foreshadows the sense of cultural appropriation discussed later in this chapter.
2. Maori for "people of the land."
3. Maori word for New Zealanders of European origin.
4. Moana Jackson is a Maori kaumatua. See http://www.indigenous-peace-conference-2008.ac.nz/profiles/Moana_Jackson.html for a detailed biography. Moana made this comment at a conference on teaching and learning, speaking about Maori knowledge.
5. Differentiation by discipline was not possible because, while some disciplines were represented by a number of participants (e.g. physics, ecology, and chemistry), the majority of disciplines in the sample included only 1–2 participants. A larger and more intentional sample would be needed for differentiation of this nature.
6. Although Cavendish (1731–1810) did publish some of his work, and attended meetings of the Royal Society, he was notoriously shy and avoided publication or even discussing his work with his peers. *Cavendish: The Experimental Life*, Christa Jungnickel and Russell McCormmach, Bucknell University Press, 1999.
7. Refering to a PowerPoint presentation by Raymond Boxman, 2004: *How to write a good paper.* See http://www.eng.tau.ac.il/~boxman/.
8. Gopen, G. D. & Swan, J. A. (1990). The science of scientific writing. American Scientist, 78(6), 550–558.
9. Anne Lamott (1995) *Bird by bird: Some instructions on writing and life.* New York: Pantheon Books.
10. See http://www.nature.com/news/specials/women/index.html.

REFERENCES

Alexander, P. A. (2003). The development of expertise: The journey from acclimation to proficiency. *Educational Researcher, 32*(8), 10–14.
Alexander, P. A. (2005). The path to competence: A lifespan developmental perspective on reading. *Journal of Literacy Research, 37*(4), 413–436.
Alexander, P. A. (2011a) Can we get from there to here? *Educational Researcher, 32*(8), 3–4.
Alexander, P. A. (2011b). The development of expertise: The journey from acclimation to proficiency. *Educational Researcher, 32*(8), 10–14.
Alger, C. L. (2009). Secondary teachers' conceptual metaphors of teaching and learning: Changes over the career span. *Teaching and Teacher Education, 25*(5), 743–751.
Artemeva, N. (1998). The writing consultant as cultural interpreter: Bridging cultural perspectives on the genre of the periodic engineering report. *Technical Communication Quarterly, 7*(3), 285–299.
Atkinson, P. (1997). Narrative turn or blind alley? *Qualitative Health Research, 7*, 325–43.
Austin, A. E. (2009). Cognitive apprenticeship theory and its implications for doctoral education: A case study example from a doctoral program in higher and adult education. *International Journal for Academic Development, 14*(3), 173–183.
Austin, A. E. & McDaniels, M. (2006). Preparing the professoriate of the future: Graduate student socialisation for faculty roles. In J. C. Smart (Ed.), *Higher Education: Handbook of theory and research* (Vol. 21, pp. 397–456). Netherlands: Springer.
Baerger, D. & McAdams, D. (1999). Life story coherence and its relation to psychological well-being. *Narrative Inquiry, 9*, 69–96.
Bartholomae, D. (1985). Inventing the university. In M. Rose (Ed.), *When a writer can't write: Studies in writer's block and other composing process problems.* New York, NY: Guildford.
Bayer, T., Curto, K. & Kriley, C. (2005). Acquiring expertise in discipline-specific discourse: An interdisciplinary exercise in learning to speak biology. *Across the Disciplines, 2*, 1–25. Retrieved from http://wac.colostate.edu/atd/articles/bayer_curto_kriley2005.cfm.
Bayer, T. & Curto, K. A. (2012). Learning to tell what you know: A communication intervention for biology students. In E. Lee Yook & W. Atkins-Sayre (Eds.), *Communication centers and oral communication programs in higher education: Advantages, challenges, and new directions* (pp. 113–130). Plymouth, UK: Lexington Books.
Bazerman, C. (1985). Physicists reading physics schema-laden purposes and purpose-laden schema. *Written Communication, 2*(1), 3–23.
Bazerman, C. (1988). *Shaping written knowledge.* Madison, WI: University of Wisconsin Press.
Bazerman, C. (1998). The production of technology and the production of human meaning. *Journal of Business and Technical Communication, 12,* 381–187.

References

Bazerman, C. (1992). From cultural criticism to disciplinary participation: Living with powerful words. In A. Herrington & C. Moran (Eds.), *Writing, teaching and learning in the disciplines* (pp. 61–68). New York, NY: Modern Language Association of America.

Bazerman, C. & Russell, D. R. (Eds.). (1994). *Landmark essays on writing across the curriculum*. Davis, CA: Hermagoras Press.

Beaufort, A. (2004). Developmental gains of a history major: A case for building a theory of disciplinary writing expertise. *Research in the Teaching of English*, 39(2), 136–185.

Beaufort, A. (2007). *College writing and beyond: A new framework for university writing instructions*. Boulder: University Press of Colorado.

Benner, P. (1984). *From novice to expert: Excellence and power in clinical nursing practice*. Reading, MA: Addison-Wesley.

Benner, P. (2004). Using the Dreyfus model of skill acquisition to describe and interpret skill acquisition and clinical judgement in nursing practice and education. *Bulletin of Science, Technology and Society*, 24(3), 189–199.

Bent, M., Gannon-Leary, P. & Webb, J. (2007). Information literacy in a researcher's learning life: the seven ages of research. *New Review of Information Networking*, 13(2), 81–99.

Bereiter, C. & Scardamalia, M. (1987). *The psychology of written composition*. Hillsdale, NJ: Erlbaum.

Bereiter, C. & Scardamalia, M. (1993). *Surpassing ourselves*. Peru, IL: Open Court.

Bishop, W. & Ostrom, H. (1997) *Genre and writing: Issues, arguments and alternatives*. Portsmouth, NH: Boynton/Cook.

Bleakley, A. (2005). Stories as data, data as stories: Making sense of narrative inquiry in clinical education. *Medical Education, 39*, 534–40.

Bentley, J. T. & Adamson, R. (2003). Gender differences in the careers of academic scientists and engineers: A literature review. Arlington, VA: National Science Foundation.

Blakeslee, A. (1997). Activity, context, interaction, and authority: Learning to write scientific papers in situ. *Journal of Business and Technical Communication*, 11(2), 125–169.

Bloom, B. S. (1985). Generalizations about talent development. In B. S. Bloom (Ed.), *Developing talent in young people* (pp. 507–549). New York: Ballentine Books

Blum, D., Knudson, M. & Henig, R. M. (2006). *A field guide for scientific writers* (2nd ed.). New York, NY: OUP.

Boice, R. (1994). *How writers journey to comfort and fluency: A psychological adventure*. Westport, CT: Praeger, Greenwood Publishing Group.

Brady, K. & Winn, T. (2014). Using metaphors to investigate pre-service primary teachers' attitudes towards mathematics. *Double Helix, 2*. Retrieved from http://qudoublehelixjournal.org/index.php/dh/article/view/37/151.

Brammer, C., Amare, N. & Campbell, K. S. (2008). Culture shock: Teaching writing within interdisciplinary contact zones. *Across the Disciplines*, 5(5). Retrieved from http://wac.colostate.edu/atd/articles/brammeretal2008.cfm.

Bransford, J. (2004). *Thoughts on adaptive expertise.* Retrieved from http://204.200.153.100/ebeling/AdaptiveExpertise.pdf.

Bressler, D. (2014). Better than business-as-usual: Improving scientific practices during discourse and writing by playing a collaborative mystery game. *Double Helix, 2.* Retrieved from http://qudoublehelixjournal.org/index.php/dh/article/view/36/155.

Brieger, K. & Bromley, P. (2014). A model for facilitating peer review in the STEM disciplines: A case study of peer review workshops supporting student writing in introductory biology courses. *Double Helix, 2.* Retrieved from http://qudoublehelixjournal.org/index.php/dh/article/view/26/152.

Brown, B. A., Reveles, J. M. & Kelly, G. J. (2005). Scientific literacy and discursive identity: A theoretical framework for understanding science learning. *Science Education, 89,* 779–802.

Brownell, S. E., Price, J. V. & Steinman, L. (2013). Science Communication to the General Public: Why We Need to Teach Undergraduate and Graduate Students this Skill as Part of Their Formal Scientific Training. *Journal of Undergraduate Neuroscience Education, 12*(1), E6–E10.

Burton, L. & Morgan, C. (2000). Mathematicians writing. *Journal for Research in Mathematics Education, 31*(4), 429–453.

Carter, M. (1990). The idea of expertise: An exploration of cognitive and social dimensions of writing. *College Composition and Communication, 41,* 265–286.

Carter, M. (2007). Ways of knowing, doing, and writing in the disciplines. *College Composition and Communication, 58*(3), 385–418.

Chamberlin, S. A. (2010). A review of instruments created to assess affect in mathematics. *Journal of Mathematics Education, 3*(1), 167–182.

Chinn, P. W. U. & Hilgers, T. L. (2000). From corrector to collaborator: The range of instructor roles in writing-based natural and applied science classes. *Journal of Research in Science Teaching, 37,* 3–25.

Choi, A., Notebaert, A., Diaz, J. & Hand, B. (2010). Examining Arguments Generated by Year 5, 7, and 10 Students in Science Classrooms. *Research in Science Education, 40*(2), 149–169.

Collins, A., Brown, J. S. & Newman, S. E. (1989). Cognitive apprenticeship: Teaching the crafts of reading, writing, and mathematics. *Knowing, learning, and instruction: Essays in honor of Robert Glaser, 18,* 32–42.

Corkery, C. (2005). Literacy narratives and confidence building in the writing classroom. *Journal of Basic Writing, 24*(1) 48–67.

Couture, B. (1999). Modelling and emulating: Rethinking agency in the writing process. In T. Kent (Ed.), *Post-process theory: Beyond the writing-process paradigm* (pp. 30–48). Southern Illinois UP: Carbondale.

Cummings, J. (2009). The doctoral experience in science: Challenging the current orthodoxy. *British Educational Research Journal, 35*(6), 977–890.

Cuthbert, D. & Spark, C. (2008). Getting a GRiP: Examining the outcomes of a pilot program to support graduate research students in writing for publication. *Studies in Higher Education, 33*(1), 77–88.

References

Dall'Alba, G. & Sandberg, J. (2006). Unveiling professional development: A critical review of the stages models. *Review of Educational Research, 76,* 383–412.

Daley, B. J. (1999). Novice to expert: An exploration of how professionals learn. *Adult Education Quarterly, 49,* 133–148.

Day, R. A. & Gastel, B. (2006). *How to write and publish a scientific paper.* Westport, CT: Greenwood.

Deane, M. & O'Neill, P. (2011). *Writing in the Disciplines.* New York, NY: Palgrave Macmillan.

Ding, H. (2008). The use of cognitive and social apprenticeship to teach a disciplinary genre initiation of graduate students into NIH grant writing. *Written Communication, 25*(1), 3–52.

Donald, A. (2013). *Mentoring: Getting personal?* Retrieved from http://occamstype writer.org/athenedonald/2013/07/30/mentoring-getting-personal/.

Doody, S. (2015). "Everything is in the lab book": The role of the lab book genre in writing, knowledge-making, and identity construction in academic medical physics labs. Unpublished master's thesis, Carleton University, Ottawa, Ontario.

Dorner, D. & Scholkopf, J. (1991). Controlling complex systems, or expertise as "grandmother's know-how." In Ericsson, K. A. & Smith, J. (Eds.), *Towards a general theory of expertise: Prospects and limits* (pp. 218–239). Cambridge, UK: CUP.

Dreyfus, H. L. (2004). The five-stage model of adult skill acquisition. *Bulletin of Science, Technology and Society, 24*(3), 177–181.

Dreyfus, Hubert L. & Dreyfus, Stuart E. (2005). Peripheral vision: Expertise in real world contexts. *Organisation Studies, 26*(5), 779–792.

Driscoll, D. L. (2011). Connected, disconnected, or uncertain: Student attitudes about future writing contexts and perceptions of transfer from first year writing to the disciplines. *Across the Disciplines, 8*(2). Retrieved from http://wac.colostate.edu/atd /articles/driscoll2011/.

Driver, R., Leach, J., Millar, R. & Scott, P. (1996). *Young people's images of science.* Philadelphia, PA: Open University Press.

Dunbar, K. (1996). How scientists think: Online creativity and conceptual change in science. In T. B. Ward, S. M. Smith & S. Vaid (Eds.), *Conceptual structures and processes: Emergence, discovery and Change* (pp. 461–493). Washington DC: APA Press.

Dweck, C. (2006). Is math a gift? Beliefs that put females at risk. In S. J. Ceci. & W. M. Williams (Eds.), *Why aren't more women in science? Top researchers debate the evidence* (pp. 47–55). Washington DC: American Psychological Association.

Ellis, C. (2004). *The ethnographic I.* Lanham, MD: AltaMira Press.

Emerson, L. (2012) An investigation of the senior academic scientist as writer in Australasian universities. In C. Bazerman et al. (Eds.), *International Advances in Writing Research: Cultures, Places, Measures* (pp. 355–372). Fort Collins, CO: The WAC Clearinghouse and Parlor Press. Retrieved from http://wac.colostate.edu /books/wrab2011/.

Emerson, L. (Spring 2004). The WAC Matrix: Institutional requirements for nurturing a team-based WAC program. *WPA: Writing Program Administration, 27*(3), 53–68.

Emerson, L., MacKay, B. R., MacKay, M. B. & Funnell, K. A. (2002). Writing in a New Zealand tertiary context: WAC and action research. In S. McLeod (Ed.), *Language and Languages special Edition: WAC in an international context*, 5, 3, 110–133.

Ericsson, K. A. (2004). Deliberate Practice and the Acquisition and Maintenance of Expert Performance in Medicine and Related Domains. *Academic Medicine*, 79(10), 570–581.

Ericsson, K. A., Krampe, R. T. & Tesch-Romer, C. (1993). The role of deliberate practice in the acquisition of expert performance. *Psychological Review*, 100, 363–406.

Ericsson, K. A., Prietula, M. & Cokely, E. (2007). The making of an expert. *Harvard Business Review*. Retrieved from http://hbr.org/2007/07/the-making-of-an-expert/ar/1.

Florence, M. K. & Yore, L. D. (2004). Learning to write like a scientist: A study of the enculturation of novice scientists into expert discourse communities by co-authoring research reports. *Journal of Research in Science Teaching*, 41, 637–668.

Fox, M. F. & Faver, C. A. (1985). Men, women, and publication productivity: Patterns among social work academics. *The Sociological Quarterly*, 26, 537–549.

Frank, A. (1995). *The wounded storyteller: body, illness, and ethics*. Chicago: Chicago University Press.

Frank, A. (2000). The standpoint of storyteller. *Qualitative Health Research*, 10, 354–65.

Gardner, S. K. (2009). *The development of doctoral students: Phases of challenge and support*. San Francisco: Jossey-Bass.

Gardner, S. K. (2007). "I heard it through the grapevine": Doctoral student socialization in chemistry and history. *Higher Education*, 54, 823–740.

Gardner, S. K. (2010). Contrasting the socialization experiences of doctoral students in high-and low-completing departments: A qualitative analysis of disciplinary contexts at one institution. *The Journal of Higher Education*, 81(1), 61–81.

Gee, J. P. (2005). The new literacy studies: From "socially situated" to the work. *Situated Literacies: Reading and Writing in Context*, 2, 177–194.

Geisler, C. (1994). Academic Literacy and the Nature of Expertise: Reading, Writing, and Knowing in Academic Philosophy. New York, NY: Lawrence Erlbaum Associates.

Golde, C. M. (2005). The role of the department and discipline in doctoral student attrition: Lessons from four departments. *The Journal of Higher Education*, 76, 669–700.

Golde, C. M. (2010). Entering different worlds: Socialisation into disciplinary communities. In S. K. Gardner & P. Mendoza (Eds.), *On becoming a scholar: Socialization and development in doctoral education* (pp. 79–95). Sterling, VA: Stylus.

Gordin, M. D. (2015). *Scientific Babel: How science was done before and after global English*. Chicago, IL: University of Chicago Press.

Grant, B. M. (2006). Writing in the company of other women: Exceeding the boundaries. *Studies in Higher Education*, 31(4), 483–495.

Grant, B. & Knowles, S. (2000). Flights of imagination: Academic women be(com)ing writers. *International Journal for Academic Development*, 5(1), 6–19.

Graves, H. (2005). *Rhetoric in(to) science: Style as invention in inquiry*. Cresskill, NJ: Hampton Press.

Greenwood, M. R. C. & Riordan, D. G. (2001). Civic scientist/Civic Duty. *Science Communication*, 23, 28–40.

References

Hand, B. M., Prain, V. & Yore, L. D. (2001). Sequential writing tasks' influence on science learning. In P. Tynjala, L. Mason & K. Lonka (Eds.), *Writing as a learning tool: Integrating theory and practice* (pp. 105–129). Dordrecht, the Netherlands: Kluwer.

Harding, P. & Hare, W. (2000). Portraying science accurately in classrooms: Emphasizing open-mindedness rather than relativism. *Journal of Research in Science Teaching, 37*(3), 225–292.

Hartley, J. & Branthwaite, A. (1989). The psychologist as wordsmith: A questionnaire study of the writing strategies of productive British psychologists. *Higher Education, 18,* 423–452.

Hartley, J. & Knapper, C. K. (1984). Academics and their writing, *Studies in Higher Education, 9,* 151–167.

Hatano, G. & Inagaki, K. (1986). Two courses of expertise. In H. Stevenshon, J. Azuma & Hakuta (Eds.), *Child development and education in Japan* (pp. 262–272). New York, NY: W.H. Freeman & Co.

Harvey, M., Mishler, E., Kroenen, K. & Harney, P. (2000). In the aftermath of sexual abuse: Making and remaking meaning in narratives of trauma and recovery. *Narrative Inquiry, 10,* 291–311.

Hidi, S. (1990). Interest and its contribution as a mental resource for learning. *Review of Educational Research, 60,* 549–571.

Hill, C., Corbett, C. & Rose, A. (2010). *Why so few? Women in science, technology, engineering and mathematics.* Sponsored by the American Association of University Women. Retrieved from http://www.aauw.org/research/why-so-few/.

Hodson, D. (1998). Is this really what scientists do? Seeking a more authentic science and beyond the school laboratory. In J. Wellington (Ed.), *Practical work in school science: Which way now?* (pp. 93–108). London: Routledge.

Holbrook, J. & Rannikmae, M. (2007). The nature of science education for enhancing scientific literacy. *International Journal of Science Education, 29*(11), 1347–1362.

Holstein, J. & Gubrium, J. (2000). *The self we live by: Narrative identity in a postmodern world.* Oxford, UK: OUP.

Holyoak, K. J. (1991). Symbolic connectionism: Toward third-generation theories of expertise. In Ericsson, K. A. & Smith, J. (Eds.), *Towards a general theory of expertise: Prospects and limits* (pp. 301–335). Cambridge, UK: CUP.

Jacoby, S. & Gonzales, P. (1991). The constitution of the expert-novice in scientific discourse. *Issues in Applied Linguistics, 2,* 149–181.

Jones, J. E. & Preusz, G. C. (1993). Attitudinal factors associated with individual factor research productivity. *Perceptual and Motor Skills, 76,* 1191–1198.

Kagan, D. (1990). Ways of evaluating teacher cognition: Inferences concerning the goldilocks principle. *Review of Educational Research, 60,* 419–16.

Kamler, B. (2008). Rethinking doctoral publication processes: Writing from and beyond the thesis. *Studies in Higher Education, 33*(3), 284–294.

Kelly, G. J. (2007). Discourse in science classrooms. In N. G. Lederman & S. K. Abell (Eds.). *Handbook of research on science education* (pp. 443–469). New York, NY: Routledge.

Keys, C. W. (1999). Revitalizing instruction in scientific genres: Connecting knowledge production in the writing to learn in science. *Science Education, 83*, 115–130.

Knisely, K. (2005). *A student handbook for writing in biology* (2nd ed.). Sunderland, MA: Sinauer Associates.

Krauss, L. (2009). An update on C. P. Snow's "two cultures." *Scientific American.* Retrieved from http://www.scientificamerican.com/article/an-update-on-cp-snows-two-cultures/.

Lajolie, Susanne P. (2003). Transitions and trajectories for studies of expertise. *Educational Researcher, 32*(8), 21–25.

Leavy, A. M., McSorley, F. A. & Boté, L. A. (2007). An examination of what metaphor construction reveals about the evolution of preservice teachers' beliefs about teaching and learning. *Teaching and Teacher Education, 23*(7), 1217–1233.

Lee, A. & Aitchison, C. (2009). Writing for the doctorate and beyond. In D. Bould & A. Lee (Eds.), *Changing practices of doctoral education* (pp. 87–99). New York, NY: Routledge.

Leshner, A. I. (2003). Public engagement with science. *Science, 299*, 977.

Lerner, N. (2007). Laboratory lessons for writing and science. *Written Communication, 24*(3), 191–222.

Leydens, J. A. (2008). Novice and insider perspectives on academic and workplace writing: Toward a continuum of rhetorical awareness. *IEEE Transactions on Professional Communication, 51*(3), 242–263.

Loudon, M. (1992). *Unveiled: Nuns talking.* London, UK: Random House.

Maher, D., Seaton, L., McCullen, C., Fitzgerald, T., Otsuji, E. & Lee, A. (2008). "Becoming and being writers": The experiences of doctoral students in writing groups. *Studies in Continuing Education, 30*(3), 263–275.

Maher, M. A., Timmerman, B. C., Feldon, D. F. & Strickland, D. (2013). Factors affecting the occurrence of faculty-doctoral student co-authorship. *The Journal of Higher Education, 84*(1), 121–143

Maher, M. A., Feldon, D. F., Timmerman, B. E. & Chao, J. (2014). Faculty perceptions of common challenges encountered by novice doctoral writers. *Higher Education Research and Development, 33*(4), 699–711.

Martin, Laura Jane (August, 2012). Scientists as writers. *Scientific American.* Retrieved from http://blogs.scientificamerican.com/guest-blog/2012/08/15/scientists-as-writers/.

Martinez, M. A., Sauleda, N. & Huber, G. L. (2001). Metaphors as blueprints of thinking about teaching and learning. *Teaching and Teacher Education 17*, 965–977.

Mayrath, M. & Robinson, D. H. (2005). Publishing, Reviewing, and Editing in Educational Psychology Journals: Comments from Editors in 1996 and 2004. *American Psychological Association, Division, 15.*

McLeod, S. H., and Soven, M. (Eds.). (2000). *Writing Across the Curriculum: A Guide to Developing Programs.* Fort Collins, CO: WAC Clearinghouse. Retrieved from http://wac.colostate.edu/books/mcleod_soven/.

References

Melzer, D. (2014). The connected curriculum: Designing a vertical transfer writing curriculum. *The WAC Journal, 25*, 78. Retrieved from http://wac.colostate.edu/journal/vol25/melzer.pdf.

Mishler, E. G. (1995). Research interviewing: A typology. *Journal of Narrative Life History, 5*, 87–123.

Montgomery, S. L. (2013). *Does science need a global language?: English and the future of research*. Chicago, IL: University of Chicago Press.

Montgomery, S. L. (1996). *The scientific voice*. New York, NY: Guilford Press.

Morss, K. & Murray, R. (2001). Researching academic writing within a structured programme: Insights and outcomes. *Studies in Higher Education, 26*(1), 35–52.

Mullin, J. A. (2008). Interdisciplinary work as professional development changing the culture of teaching. *Pedagogy, 8*(3), 495–508.

Mutnick, D. (1998). Rethinking the personal narrative: Life-writing and composition pedagogy. In C. Farris & Anson, C. M. (Eds.), *Under construction: Working at the intersections of composition theory, research, and practice* (pp. 79–92). Logan: Utah State UP.

Norris, S. P. & Phillips, L. M. (2003). How literacy is its fundamental sense is critical to scientific literacy, *Science Education, 87*, 222–240.

O'Gorman, E., Ingle, J. & Mitchell, S. (2014). Promoting Student Participation in Scientific Research: An Undergraduate Course in Global Change Biology. *Double Helix, 2*. Retrieved from http://qudoublehelixjournal.org/index.php/dh/issue/view/2/showToc.

Olson, R. (2009). *Don't be such a scientist*. Washington, DC: Island Press.

O'Shaughnessy, R., Dallos, R. & Gough, A. (2012). A narrative exploration of the lives of women who experience anorexia nervosa. *Qualitative Research in Psychology, 10*(1), 42–62.

Paré, A. (2011). Speaking of writing: Supervisory feedback and the dissertation. In L. McAlpine & C. Amundsen (Eds.). *Supporting the doctoral process: research-based strategies* (pp. 59–74). New York, NY: Springer.

Paretti, M. C. & McNair, L. D. (2008). Introduction to the special issue on communication in engineering curricula: Mapping the landscape. *IEEE Transactions on Professional Communication, 51*(3), 238–241.

Penrose, A. M. & Katz, S. B. (2010). Writing in the sciences: Exploring conventions of scientific discourse (3rd edition). Boston, MA: Allyn and Bacon.

Pittam, G., Elander, J., Lusher, J., Fox, P. & Payne, N. (2009). Student beliefs and attitudes about authorial identity in academic writing. *Studies in Higher Education, 34*(2), 153–170.

Poe, M., Lerner, N. & Craig, J. (2010). *Learning to communicate in science and engineering*. Boston, MA: MIT Press.

Prain, V. & Hand, B. (1999). Students' perceptions of writing for learning in in secondary school science, *Science Education, 83*, 151–162.

Prain, V. (2006). Learning from writing in secondary science: Some theoretical and practical implications, *International Journal of Science Education, 28*(2–3), 179–201.

Pratt, M. L. (1991). Arts of the contact zone. *Profession, 91*, 33–40.

Raymond, J. (2013). Sexist attitudes: Most of us are biased. *Nature, 495,* 33–34

Reed, I., Pearlman, S. J., Millard, C. & Carillo, D. (2014). Peer assessment of writing and critical thinking in STEM: Insights into student and faculty perceptions and practices. *Double Helix, 2,* Retrieved from http://qudoublehelixjournal.org/index.php/dh/issue/view/2/showToc.

Reynolds, J., Smith, R., Moskovitz, C. & Sayle, A. (2009). BioTAP: A systematic approach to teaching scientific writing and evaluating undergraduate theses. *Bioscience, 59*(10), 896–903.

Reynolds, J. A., Thaiss, C., Katkin, W. & Thompson, R. J. (2012). Writing-to-learn in undergraduate science education: A community-based, conceptually driven approach. *CBE-Life Sciences Education, 11*(1), 17–25.

Rich, J. A. & Grey, C. M. (2003). Qualitative research on trauma surgery: Getting beyond the numbers. *World Journal of Surgery, 27,* 957–961.

Riel, M. (2000). Education in the 21st Century: Just-in-time learning or learning communities. *Presentation at the Fourth Annual Conference for the Emirates Center for Strategic Studies and Research,* Abu Dhabi. Retrieved from https://www.researchgate.net/publication/258698171_Education_in_the_21st_Century_Just-in-Time_Learning_or_Learning_Communities.

Rivard, L. O. P. (1994). A review of writing to learn in science: Implications for practice and research. *Journal of Research in Science Teaching, 31*(9), 969–983.

Rodgers, R. & Rodgers, N. (1999). The sacred spark of academic research. *Journal of Public Administration Research & Theory, 9*(3), 473–492.

Rodriguez, I., Goertzen, R. M., Brewe, E. & Kramer, L. (2012). Communicating scientific ideas: One element of physics expertise. *2011 Physics Education Research Conference, 1413*(1), 319–322.

Roth, W. M. & Lee, S. (2002). Scientific literacy as collective praxis. *Public Understanding of Science, 11*(1), 33–56.

Rowell, P. A. (1997). Learning in school science: The promises and practices of writing. *Studies in Science Education, 30,* 19–56.

Russell, C. B. & Weaver, G. (2008). Student perceptions of the purpose and functions of the laboratory in science: A grounded theory study. *International Journal for the Scholarship of Teaching and Learning 2*(2). Retrieved from http://digitalcommons.georgiasouthern.edu/cgi/viewcontent.cgi?article=1127&context=ij-sotl.

Russell, D. R. (1991). *Writing in the academic disciplines, 1870–1990.* Carbondale: Southern Illinois University Press.

Science for all. (2013.) Editorial. *Nature, 495,* 5.

Science Media Centre. (2015). Going public at the New Zealand Association of Scientists annual conference. Retrieved from http://www.sciencemediacentre.co.nz/2015/04/16/going-public-at-the-new-zealand-association-of-scientists-annual-conference/.

Segal, J., Pare, A., Brent, D. & Vipond, D. (1998). The researcher as missionary: Problems with rhetoric and reform in the disciplines. *College Composition and Communication, 50*(1), 71–90.

References

Shah, J., Shah, A. & Pietrobon, R. (2009). Scientific writing of novice researchers: What difficulties and encouragements do they encounter? *Academic Medicine, 84,* 511–516.

Shanahan, C. (2004). Teaching science through literacy. In T. L. Jetton & J. A. Dole (Eds.), *Adolescent Literacy Research and Practice* (pp.75–93). New York, NY: The Guilford Press.

Shaughnessy, M. (1979). *Errors and expectations: A guide for the teacher of basic writing.* New York, NY: Oxford University Press.

Slinn, E.W. (1991). *The discourse of self in Victorian poetry.* University of Virginia Press.

Smith, B. & Sparkes, A.C. (2006). Narrative inquiry in psychology: Exploring the tensions within. *Qualitative Research in Psychology, 3,* (3), 169–192.

Smith, B. & Sparkes, A.C (2008). Narrative and its potential contribution to disability studies. *Disability & Society, 23*(1), 17–28.

Smith, K. A., Sheppard, S. D., Johnson, D. W. & Johnson, R. T. (2005). Pedagogies of engagement: classroom based practices. *Journal of Engineering Education, 94*(1), 87–101.

Soliday, M. (1994). Translating self and difference through literacy narratives. *College English, 56*(5), 511–526.

Soliday, M. (2011). *Everyday genres: Writing assignments across the disciplines.* Carbondale: SIU Press.

Stanford, J. S. & Duwel, L. E. (2013). Engaging biology undergraduates in the scientific process through writing a theoretical research proposal. *Bioscene: Journal of College Biology Teaching, 39*(2), 17–24.

Starke-Meyerring, D. (2011). The paradox of writing in doctoral education: Student experiences. In L. McAlpine & C. Amundson (Eds.), *Doctoral education: Research-based strategies for doctoral students, supervisors and administrators* (pp. 75–95). New York, NY: Springer.

Stephens, C. & Breheny, M. (2013). Narrative analysis in psychology research: An integrated approach to interpreting stories. *Qualitative Research in Psychology, 10,* (1), 14–27.

Sweeting, K. (2011) *Early years teachers' attitudes towards mathematics.* Unpublished master's by research thesis, Queensland University of Technology. Retrieved from http://eprints.qut.edu.au/46123/.

Sweitzer, V. (2009). Towards a theory of doctoral student professional identity development: A developmental networks approach. *The Journal of Higher Education, 80*(1), 1–33.

Tapia, M. & Marsh, G. E. (2004). An instrument to measure mathematics attitudes. *Academic Exchange Quarterly, 8,* 16–21.

Terkel, S. (Ed.). (1974). *Working: People talk about what they do all day and how they feel about what they do.* New York, NY: The New Press.

Terkel, S. (1992). *Race: How blacks and whites think and feel about the American obsession.* New York, NY: The New Press.

Terkel, S. (1997). *The good war: An oral history of World War II.* New York, NY: The New Press.

Terkel, S. (2014). *Will the circle be unbroken?: Reflections on death, rebirth, and hunger for a faith*. New York, NY: The New Press

Thaiss, C. (2012). Origins, aims, and uses of writing programs worldwide: profiles of academic writing in many places. In C. J. Thaiss, G. Bräuer, P. Carlino, L. Ganobcsik-Williams & A. Sinha (Eds.), *Writing programs worldwide: Profiles of academic writing in many places* (pp. 5–22). Fort Collins, CO: The WAC Clearinghouse and Parlor Press. Retrieved from http://wac.colostate.edu/books/wpww/.

Tobin, K. & Tippens, D. J. (1996). Metaphors as seeds for conceptual change. *Science Teacher Education 80*(6), 711–30.

VanSledright, B. & Alexander, P. A. (2002). Historical knowledge, thinking, and beliefs: Evaluation component of the Corps of Historical Educational Researcher 14 Discovery Project (#S215X010242). Washington, DC: United States Department of Education.

Vargas, S. K. & Hanstedt, P. (2014). Exploring Alternatives in the Teaching of Lab Report Writing: Deepening Student Learning Through a Portfolio Approach, *Double Helix, 2*. Retrieved from http://qudoublehelixjournal.org/index.php/dh/article/view/28/145.

Veronikas, S. & Shaughnessy, M. F. (2005). An interview with Richard Mayer. *Educational Psychology Review, 17*(2), 179–189.

Verschaffel, L., Luwel, K., Torbeyns, J. & Van Dooren, W. (2011). Analyzing and developing strategy flexibility in mathematics education. In J. Elen, E. Stahl, R. Bromme & G. Clarebout (Eds.), *Links between beliefs and cognitive flexibility* (pp. 175–197). New York, NY: Springer.

White, A. L., Way, J., Perry, B. & Southwell, B. (2005). Mathematical attitudes, beliefs and achievement in primary pre-service mathematics teacher education. *Mathematics Teacher Education and Development, 7*, 33–52.

Yancey, K. B. & Huot, B. (Eds.) (1997). *Assessing writing across the curriculum*. Greenwich, CN: Ablex.

Yore, L. D., Hand, B. M. & Prain, V. (2002). Scientists as writers. *Science Education, 86*, 672–692.

Yore, L. D., Bisanz, G. L. & Hand, B.M. (2003). Examining the literacy component of science literacy: 25 years of language arts and science research. *International Journal of Science Education, 25*, 689–725.

Yore, L. D., Hand, B. M. & Florence, M. K. (2004). Scientists' views of science, models of writing, and science writing practices. *Journal of Research in Science Teaching, 41*, 338–369.

Yore, L. D., Florence, M. K., Pearson, T. W. & Weaver, A. J. (2006). Written discourse in scientific communities: A conversation with two scientists about their views of science, use of language, role of writing in doing science, and compatibility between their epistemic views and language. *International Journal of Science Education, 28* (2–3), 109–141.

APPENDIX A
QUESTIONNAIRE FOR THE SENIOR AND EMERGING SCIENTISTS

A. BACKGROUND

1. What is your field of research? (both broad and specific)
2. In what country/countries did you go to school (primary and secondary)?
3. In what country/countries did you complete your undergraduate qualification?
4. In what country/countries did you complete your post graduate qualifications?
5. How is your work funded? If you are funded by an outside agency, what are the expectations around research outputs?

B. BREADTH OF WRITING

1. In the last 6 months, have you conducted any of the following activities:
    - Written teaching materials
    - Brainstormed or made rough notes for a new project
    - Emailed a colleague/peer to discuss an idea for a new project
    - Taken notes or looked at notes made during a group discussion of a new project
    - Drafted a research proposal
    - Edited a research proposal (for a project you will be involved with)
    - Edited a research proposal (for a project you won't be involved with)
    - Written up field or research notes
    - Emailed a colleague/peer about how to solve a particular problem with a research project
    - Drafted a scientific paper for publication
    - Re-drafted a paper which has been written by a colleague which will include your name as part of the research team
    - Edited a scientific paper for publication

Appendix A

- Peer reviewed a scientific paper for a colleague (i.e. prior to submission)
- Peer reviewed a scientific paper for a journal (i.e. as part of the formal peer review process)
- Edited a journal or been part of an editing team
- Drafted a paper for a popular science journal
- Edited a paper for a popular science journal
- Written on a scientific topic on a website (e.g. blog, forum)
- Engaged in writing a piece of creative writing or creative non-fiction
- Submitted for publication a piece of creative writing or creative non-fiction
- Written or drafted a report for a professional body (e.g. regional council, government department
- Written or drafted an in-house document e.g. a report?
- Other (please specify)

2. Of the activities described above, please indicate which ones (up to 5) took up most time in your professional activities over the last 6 months.

C. RESOURCING

1. Do you ever use the services of a professional editor?
2. Does your organisation provide you with a writing guide/style manual?
3. If yes, do you use it? Do you find it helpful?
4. Have you ever attended a writing workshop provided by an organisation you work for?
5. Have you ever attended a writing workshop provided by someone or a group outside your organisation (e.g. Royal Society)?
6. Do you normally have your work peer reviewed by a colleague before it is submitted to a journal?

APPENDIX B
INTERVIEW SCHEDULE FOR SENIOR AND EMERGING SCIENTISTS

A. FOLLOW-UP QUESTIONS FROM THE QUESTIONNAIRE

1. Are there any issues you would like to expand on from the questionnaire?
2. Follow up questions that emerge from the questionnaire (if appropriate)

B. WRITING PROCESS

1. Could you describe in detail a process you went through in writing up a particular project? Could you start right from the beginning—for example, could you think right back to the conceptualisation of the project.
2. When you design a project, do you have a journal in mind from the very beginning? How do you choose the journal?
3. Do you use writing to conceptualise your project? Or do you start writing when the project is fully formed in your head? Or when the research is complete and you're writing up?
4. Are you aware of your audience during your writing? When does this become a factor—from the beginning or from the writing up stage? Does it affect how you write?
5. Have you ever written for an audience you're not very comfortable with e.g. outside your discipline or for a broad scientific audience? How did you deal with that? Did you enjoy it? How did you go about it?
6. Do you feel, when you're writing, that you are being persuasive? Do you think there is a role for persuasion in science writing?

C. ATTITUDES TO WRITING

1. Do you think you're a good scientific writer?
2. What writing do you find the easiest?
3. What do you find most difficult?
4. How do you feel about the writing you do within your science discipline?

5. What are some of the greatest hurdles to your writing? How do you feel about those hurdles?
6. Describe some successful writing you have done. How did you feel about that? Why was it a success?
7. What motivates you to keep writing within your discipline?
8. Would you say you enjoy writing?
9. What is your understanding of the role of writing science (scientific writing)?
10. What might you perceive to be the relationship of writing and science?
11. Do you learn new things when you're writing?
12. Scientific writer's lifecycle: do you agree with this model? Where do you see yourself in relation to it?
13. Would you classify yourself as someone who likes to finely prescribe a particular kind of writing (e.g. writing purely in your field to a specific audience) and do that well—or do you like new challenges, new audiences?
14. What do you think are the particular stylistic issues that are important for writing in your discipline?
15. Writing in a team. If you engage in team research how do you undertake the writing process?

D. LEARNING TO WRITE SCIENCE

1. Did you enjoy writing as a child? If you did, what did you enjoy? If you didn't, why was that?
2. Do you feel you were taught the basics of writing well at school? For example, did you leave school feeling you had a good grasp of grammar, punctuation, spelling? What about slightly higher order activities such as paragraphing or how to construct an essay?
3. Did you take any writing based courses, e.g. English or History, right through high school?
4. Did you have any training in writing science at school?
5. How, where, when did you learn to write as a scientist?
6. Was there anyone or anything that had a major influence on you as a writer of science? Was it a positive or negative influence?
7. When you look back at your development as a scientific writer, is there anything you wish you'd done differently?
8. Does the system you experienced, in becoming a writer of science, worked well? Is there any way in which you could have been better helped to become a writer of science?

E. ROLE AS A TEACHER

1. Do you think you have a role as a teacher of science writing?
2. If you do, how do you do it?
3. Do you see this as an important part of your role as a scientist? Is it a role you enjoy?

F. OTHER

1. Is there any kind of writing you'd like to do more often?
2. Do you write for pleasure? If so, what kind of writing do you engage in?
3. Do you read for pleasure? If so, what do you read?
4. Is there anything else you'd like to say about writing and science? Is there anything else we haven't asked you—about writing in general, or writing in your profession, that you think is important for us to understand in learning about how scientists become writers of science?

APPENDIX C
INTERVIEW FOR PH.D. STUDENTS

A. FOLLOW-UP QUESTIONS FROM THE QUESTIONNAIRE

1. Are there any issues you would like to expand on from the questionnaire?
2. Follow up questions that emerge from the questionnaire (if appropriate)

B. THE PH.D. EXPERIENCE

1. Tell me about your dissertation now—where are you up to and how is it going?
2. How did you get your topic—was it something you initiated, or was the Ph.D. advertised? How did you get into it and how did you find your topic?
3. Is your dissertation a standalone project or does it fit into a bigger research project? How does that work in practical terms?
4. Can you talk me through the process you've been through so far with your dissertation? What did you do first? What kind help did you get with that? How were you guided (and by whom)? What happened next?
5. Who provides you with feedback? What form does the feedback take? Does the person (or people) providing feedback rewrite or do they make suggestions for how you might rewrite? Are their directions clear? Have you ever disagreed with the feedback?
6. Do you feel confident about writing up the dissertation?
7. What has been helpful for you in writing for the dissertation?
8. What are the biggest problems you face in writing the dissertation?
9. Looking back on the process so far, is there anything you'd do differently if you were starting again?
10. Can you think of anything that is not available to you that would be helpful for you in writing up the dissertation?

C. GENERAL QUESTIONS ABOUT WRITING

1. When you're involved in a project, who decides what journal or conference it's being aimed for? Do you have a journal in mind from the very beginning? How do you choose the journal?

2. Do you use writing to conceptualise your project, or do you start writing when the project is fully formed in your head? Or when the research is complete?
3. Are you involved in collecting and recording data? If so, what kind of writing do you do at this stage?
4. Are you aware of your audience during your writing? Who IS your audience? Does it affect how you write? If so, how? Do you feel comfortable writing for this audience?
5. Have you ever written for a different kind of audience (than the dissertation)? How did you deal with that? Did you enjoy it? How did you go about it?
6. Do you feel, when you're writing up a paper or your dissertation, that you are being persuasive? Do you think there is a role for persuasion in science writing?

D. ATTITUDES TO WRITING

1. Do you think you're a good writer?
2. What writing do you find the easiest?
3. What do you find most difficult?
4. Would you say you enjoy writing?
5. What are some of the greatest hurdles to your writing? Describe some successful writing you have done. How did you feel about that? Why was it a success?
6. What is your understanding of the role of writing science (scientific writing)?
7. What might you perceive to be the relationship of writing and science?
8. Do you learn new things when you write?
9. Lifecycle question: do you agree with this model? Where do you see yourself in relation to it?
10. Would you classify yourself as someone who likes to finely prescribe a particular kind of writing (e.g. writing purely in your field to a specific audience) and do that well—or do you like new challenges, new audiences?
11. What do you think are the particular stylistic issues that are important for writing in your discipline?

E. LEARNING TO WRITE SCIENCE

1. Did you enjoy writing as a child? If you did, what did you enjoy? If you didn't, why was that?

2. Do you feel you were taught the basics of writing well at school? For example, did you leave school feeling you had a good grasp of grammar, punctuation, spelling? What about slightly higher order activities such as paragraphing or how to construct an essay?
3. Did you take any writing based courses, e.g. English or History, right through high school?
4. Did you have any training in writing science at school?
5. How, where, when did you learn to write as a scientist?
6. Is there anyone or anything who has had a major influence on you as a writer of science? Was it a positive or negative influence?
7. When you look at your development as a scientific writer, is there anything you wish you'd done differently?
8. Does the system you're experiencing, in becoming a writer of science, working well? Is there any way in which you could have been better helped to become a writer of science?

F. OTHER

1. Is there any kind of writing you'd like to do more often?
2. Do you write for pleasure? If so, what kind of writing do you engage in?
3. Do you read for pleasure? If so, what do you read?
4. Is there anything else you'd like to say about writing and science? Is there anything else we haven't asked you—about writing in general, or writing in your profession, that you think is important for us to understand in learning about how scientists become writers of science?

APPENDIX D
THE SCALE USED TO DEVELOP THE QUANTITATIVE DATA

Translating the interviews into quantitative data was a necessary process in providing useful methods of comparison across different groups (i.e. gender, different lifecycle stages and adaptive/routine/transitioning scientists). However, developing a reliable process for doing this presented difficulties. Within their interview, participants regularly contradicted themselves or expressed a point of view that modified their answer to a direct question. For example, several participants argued in a response to a direct question that scientific writing was never and should not be persuasive, but then at other points in the interview made observations which suggested that their perspective was more nuanced than this. Working with only anonymized responses to direct questions lifted from context (the best way to maximize objectivity in the assessment) would therefore have produced extremely inaccurate results in many cases.

For this reason, the variables identified in the model were converted to quantitative data based on the entire interview, from answers to direct questions (e.g. is scientific writing persuasive?), indirect questions (e.g. can you show me what you mean by 'story'?), and the interview as a whole (for example, if a participant said that scientific writing was never persuasive but later in the interview while discussing something else demonstrated evidence of seeing scientific writing as persuasive, this was used to modify the original rating). The analysis was conducted twice, four weeks apart, with the names of the participants removed, and where discrepancies were apparent between the two sets of analysis, the transcript was assessed again. While this approach sacrificed the element of anonymity in a small number of the interviews (a few interviews were so distinctive and memorable that they were easily identifiable), this was considered the most reliable approach.

Each variable was allocated a mark out of 10, using the scales provided below.

Quadrant 1: Early Experiences

Childhood Attitudes

1. Strongly Negative				5. Neutral				10. Strongly Positive

Appendix D

Childhood Experiences

1. Strongly Negative				5. Neutral				10. Strongly Positive

Quadrant 2: Learning to Write Science

Help from Advisor

1. None				5. Some Useful Support				10. Sustained and Extensive

Help from Community (e.g. lab partners, friends, family)

1. None				5. Some Useful Support				10. Sustained and Extensive

Help for Rhetorical Reading

1. None				5. Some Useful Support				10. Sustained and Extensive

Ongoing Support Post-Ph.D.

1. None				5. Some Useful Support				10. Sustained and Extensive

Quadrant 3: Attitudes

Enjoyment

1. None				5. Neutral				10. Extreme

Motivation

1. None				5. Neutral				10. Extreme

Resilience

1. None				5. Neutral				10. Extreme

Strength of Self-efficacy/Purpose

1. None				5. Neutral				10. Extreme

QUADRANT 4: BELIEFS

Function of Writing

1. Unnecessary				5. Sometimes Discovery, Sometimes Reporting				10. Is Always about Discovery and Reporting

Importance of Audience

1. None				5. Neutral				10. Extremely

Importance of persuasion

1. None				5. Neutral				10. Extremely

Beliefs about Identity/Role as a Scientist

1. Role is to move science forward	Role is about disciplinary contribution and change	Role is about disciplinary leadership	Role is about disciplinary leadership/ cross-disciplinary connection	Role is about disciplinary leadership/ reaching out to a broader audience	Role is about disciplinary leadership and/ public leadership	10. Role is to change society

www.ingramcontent.com/pod-product-compliance
Lightning Source LLC
Chambersburg PA
CBHW020404080526
44584CB00014B/1160